SpringerBriefs in Applied Sciences and Technology

SpringerBriefs present concise summaries of cutting-edge research and practical applications across a wide spectrum of fields. Featuring compact volumes of 50 to 125 pages, the series covers a range of content from professional to academic.

Typical publications can be:

- A timely report of state-of-the art methods
- An introduction to or a manual for the application of mathematical or computer techniques
- A bridge between new research results, as published in journal articles
- A snapshot of a hot or emerging topic
- An in-depth case study
- A presentation of core concepts that students must understand in order to make independent contributions

SpringerBriefs are characterized by fast, global electronic dissemination, standard publishing contracts, standardized manuscript preparation and formatting guidelines, and expedited production schedules.

On the one hand, **SpringerBriefs in Applied Sciences and Technology** are devoted to the publication of fundamentals and applications within the different classical engineering disciplines as well as in interdisciplinary fields that recently emerged between these areas. On the other hand, as the boundary separating fundamental research and applied technology is more and more dissolving, this series is particularly open to trans-disciplinary topics between fundamental science and engineering.

Indexed by EI-Compendex, SCOPUS and Springerlink.

Tajul Adli Abdul Razak · Ahmad Kamal Ismail ·
Andreas Öchsner
Editors

IT Application for Sustainable Living 2

Springer

Editors
Tajul Adli Abdul Razak
Malaysian Spanish Institute
Universiti Kuala Lumpur
Kulim, Kedah, Malaysia

Ahmad Kamal Ismail
Malaysian Spanish Institute
Universiti Kuala Lumpur
Kulim, Kedah, Malaysia

Andreas Öchsner
Faculty of Mechanical Engineering
Esslingen University of Applied Science
Esslingen am Neckar, Baden-Württemberg,
Germany

ISSN 2191-530X ISSN 2191-5318 (electronic)
SpringerBriefs in Applied Sciences and Technology
ISBN 978-3-031-93197-0 ISBN 978-3-031-93198-7 (eBook)
https://doi.org/10.1007/978-3-031-93198-7

This Springer imprint is published by the registered company Springer Nature Switzerland AG
The registered company address is: Gewerbestrasse 11, 6330 Cham, Switzerland

Preface

In today's fast-paced and competitive business landscape, industries are grappling with challenges in process optimization, energy efficiency, and transportation. Effective management of these areas is essential for boosting productivity while minimizing waste and fostering sustainable growth for future business. This collection of research and development projects highlights innovative solutions across various sectors, including electric vehicle performance optimization, smart agriculture, healthcare monitoring, and industrial maintenance. Key focuses include electric vehicle motor and battery technology advancements, driver safety systems, IoT-based vehicle surveillance, and energy-efficient charging solutions. Additionally, cutting-edge applications like Li-Fi for vehicle-to-vehicle data transfer and AI-driven auto-sorting systems demonstrate the integration of modern technology to address real-world challenges. These projects collectively aim to enhance operational efficiency, reduce environmental impact, and drive technological progress across industries.

Kulim, Malaysia	Tajul Adli Abdul Razak
Kulim, Malaysia	Ahmad Kamal Ismail
Esslingen am Neckar, Germany	Andreas Öchsner

Contents

Chapter 1
Performance Characteristics of Electric Vehicle Using Different Motor and Battery Combination

Shahril Nizam Mohamed Soid, Mohamad Sodiqul Aufa Fadillah, Khairul Shahril Shaffee, Tajul Adli Abdul Razak, and Ishak Abd. Azid

Abstract The transition of internal combustion engines to electric motors as power source is increasing in the automotive field which leads to an increase in the total production of electric vehicles. The production of EVs and the construction of charging infrastructure will be a significant problem for the EV industry. There will be a feeling of concern by the user about the battery capacity, driving distance and charging time. Therefore, the selection of the combination of motor type and battery for the EV drivetrain is very important to enhance the EV performance. This current study aims to develop an EV model with an optimized combination of motor and battery types. The EV performance was investigated in terms of torque, acceleration, distance, and state of charge. The EV model was developed using the Gamma Technologies GT-SUITE software. The performance was assessed with a different combination of motors (induction motor and permanent magnet synchronous motor) and batteries (lithium iron phosphate, lithium nickel cobalt oxide and lithium nickel cobalt aluminium oxide). From the analysis, it was found that the combination of induction motor and lithium nickel cobalt oxide had the best performance compared to others.

Keywords Electric vehicle · Battery · GT-suite · Induction motor · Permanent magnet synchronous motor · Lithium battery

---

S. N. M. Soid (✉) · M. S. A. Fadillah · K. S. Shaffee · T. A. A. Razak · I. Abd. Azid
Universiti Kuala Lumpur Malaysian Spanish Institute, Kulim, Kedah, Malaysia
e-mail: shahrilnizam@unikl.edu.my

K. S. Shaffee
e-mail: khairuls@unikl.edu.my

T. A. A. Razak
e-mail: tajuladli@unikl.edu.my

I. Abd. Azid
e-mail: ishak.abdulazid@unikl.edu.my

T. A. A. Razak et al. (eds.), *IT Application for Sustainable Living 2*,
SpringerBriefs in Applied Sciences and Technology,
https://doi.org/10.1007/978-3-031-93198-7_1

1.1 Introduction

The transportation industry is one of the major contributors to pollutant emissions, which mainly comes from the usage of conventional internal combustion engines. The petrol engine emits CO_2 while diesel engine produces soot and particulate that pollute the atmosphere. Both of petrol and diesel engines are frequently used in transportation applications, and produce emissions that cause air pollution, sickness, which has resulted in the deterioration of the urban air quality [1]. The usage of electric vehicles is becoming popular due to their zero carbon emissions and high efficiency. In the manufacture of electric vehicles, the motor and battery are two essential components. The battery-driven electric motor is an important component of providing power for electric vehicles (EVs). An electric motor is an electric machine that converts electrical energy into mechanical energy. These electric motors are often powered by inverters, which use direct current (DC) sources such as batteries and motor vehicles and alternating current (AC) sources as the influence grid. Various types of electric motors are used in EVs and four common types of electric motor include DC motors, induction motors (IM), permanent magnet synchronous motors (PMSM) and switched reluctance motors (SRM). Most researchers have conducted studies that focus more on batteries and motors [2–5]. Yildirim et al. [6] also had studied various types of electrical motor and drivetrain systems on EV. They had found that the switch reluctant motors is an appropriate motor for EVs due to its low weight and reliability.

In addition to study on EV motors, Iclodean et al. [7] had studied four different types of batteries, that are lithium ion (Li-Ion), molten salt ($Na–NiCl_2$), nickel metal hydride (Ni–MH) and lithium sulphur (Li–S) using the AVL cruise software. The results of the study have shown that in terms of energy consumption, $Na–NiCL_2$ batteries have proven to be the best choice. However, the operating temperature of the battery is increased that led to solidification of electrolyte if the vehicle is not used. EV are largely dependent on their power batteries for their performance [8]. The battery pack ultimately determines an EV's maximum range, efficiency, charging time, state of charge (SOC) and running costs. Lead acid, nickel cadmium, nickel-metal hydride, molten salt, lithium polymer and lithium-ion batteries are the six main types of batteries that have been used in EVs. Tran et al. [9] had experimentally compared the performances of three equivalent circuit model with four types of common Li-Ion batteries, namely lithium iron phosphate (LFP), lithium nickel manganese cobalt oxide (NMC), lithium manganese oxide (NCA) and lithium manganese oxide (LMO). It was found that the best-performed model for LFP and NCA was the 1RC with hysteresis ECM, while the most suited model for NMC and LMO was the 1RC ECM. Most consumer electronics and EV applications today use lithium-ion (Li-ion) batteries as their primary power source. Their high energy ratings and specific powers have contributed significantly to their success [10].

In the current research, the effect of motor and battery combination on the efficiency of an electric vehicle is investigated. This study is divided into three sections, starting with the EV model development, followed by validation of the model. The designed electric vehicle system was validated with the results of the simulation of the commercialized car.

1.2 Methodology

The EV system was modelled using GT-ISE where the design of the EV system was developed without complete auxiliary load. This design was referred to the electric vehicle template that was prepared in GT-ISE in the form of GT project map. The parts and templates in the main assembly were battery part, motor part and speed control template. The main assembly also consisted of three subassemblies which were the vehicle, motor control and braking control. Figure 1.1 shows the design of EV model main assembly.

A total of six model systems had been developed with different motor and battery combinations for this study as shown in Table 1.1. The type of battery used in this study was lithium iron phosphate (LFP), lithium nickel cobalt aluminium oxide (NCA) and lithium nickel manganese cobalt oxide (NMC). For the motor, the types of motor studied were induction motors (IM) and permanent magnet synchronous motors (PMSM).

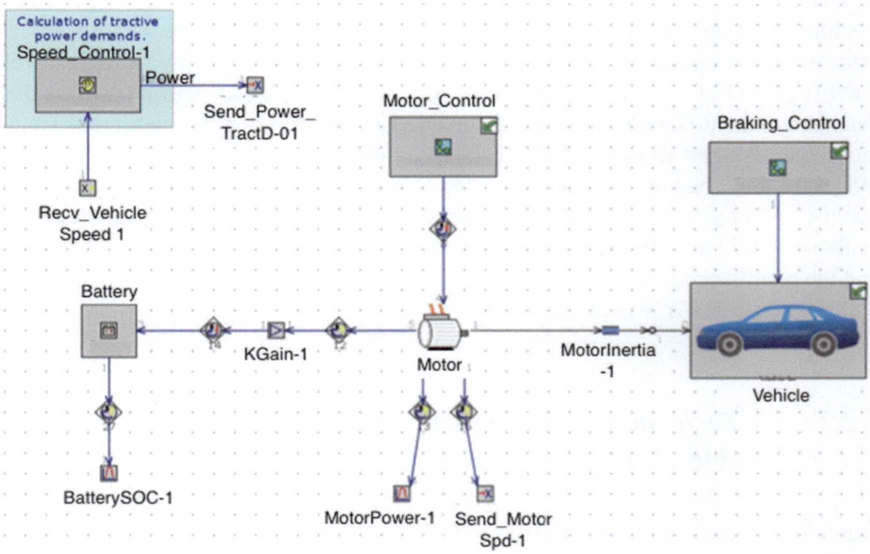

Fig. 1.1 Completed EV model main assembly

Table 1.1 Battery and motor combinations

No.	Battery type	Motor type	Combination model
1	LFP	IM	LFP IM
2	LFP	PMSM	LFP PMSM
3	NCA	IM	NCA IM
4	NCA	PMSM	NCA PMSM
5	NMC	IM	NMC IM
6	NMC	PMSM	NMC PMSM

Table 1.2 Distance travel comparison of EV model (this study) with ANL for each drive cycles

Drive cycles	EV model (km)	ANL (km)	Percentage different (%)
UDDS	11.869	11.974	0.88
HWY	16.037	16.500	2.80

1.3 Result and Discussion

1.3.1 Model Validation

For the model validation, a commercially available EV model with the same specifications which was a Nissan Leaf 2013, as tested by the Advanced Powertrain Research Facility (APRF) from the Argonne National Laboratory (ANL). The parameters that had been compared were the distance travelled by the EV after completing the simulation based on the set drive cycles. Table 1.2 shows the distance travelled by the EV model for the urban dynamometer driving Schedule (UDDS) drive cycle which is at about 11.869 km and for the highway fuel economy test (HWY) drive cycle which is at about 16.037 km. The results show the correlation between the simulation model and the results from ANL where the percentage of different for UDDS was at about 0.88% and for HWY was at about 2.8%.

1.3.2 EV Performance Analysis Based on Different Combination of Motor and Battery

Figure 1.2 shows the state of charge (SoC) comparison for each combination of battery and motor types for a duration of 5500 s running time. Each combination showed a slight reduction from the initial SoC value with the combination of NCA PMSM and NCA IM recording indicate the least reduction of SoC where the final values were 0.84 and 0.84, respectively. It is followed by the combination of NMC PMSM and LFP PMSM where each recorded a final SoC value of 0.82 and 0.80,

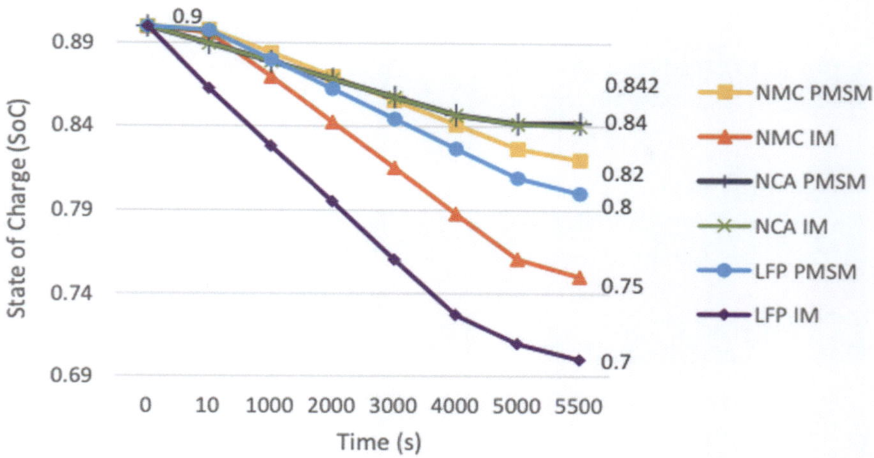

Fig. 1.2 SoC comparison for each combination of battery and motor types after 5500 s running time

respectively. The final SoC value of the NMC IM combination recorded a significant reduction with a final SoC value of 0.75. The highest reduction of SoC was recorded by the LFP IM combination with a final value of 0.70 SoC.

Figure 1.3 shows the maximum torque for each battery and motor combination. NMC IM obtained the highest maximum torque value which was about 347.62 Nm, while NCA PMSM produced the lowest maximum torque value about 83.57 Nm. It also showed that every battery with IM combination produced a higher maximum torque value compared to the battery with PMSM combination.

Figure 1.4 shows the comparison of acceleration for all battery and motor combinations. From the figure, it was found that the highest acceleration value is produced by LFP IM with an acceleration value of 2.57 m/s^2. The lowest acceleration value is 0.59 m/s^2 for NCA PMSM. The vehicle weight is also one of the factors that affect the acceleration. LFP IM a has higher acceleration value than the NCA PMSM due to the fact that the weight for the LFP IM combination is lighter compared to NCA PMSM. The total weight for the LFP IM combination is 627 kg while the weight for NCA PMSM is 631 kg which is the heaviest compared to all combinations. The same thing can be seen in the weight difference owned by NMC IM with a weight of 717 kg which is heavier than LFP IM.

Figure 1.5 shows the distance travelled by all battery and motor combinations for 5500 s running time. Each of the combination shows a gradual increase in the distance travelled with a different final value at the end of the simulation. NMC IM had produced the longest distance travelled at about 167.77 km while NCA PMSM had produced the shortest distance travelled at about 109.71 km. There were several reasons such as the NMC IM had a higher acceleration value compared to NCA PMSM. The acceleration of an EV determines how quickly it can reach a particular speed, and the time required to achieve that speed, that affected the distance travelled.

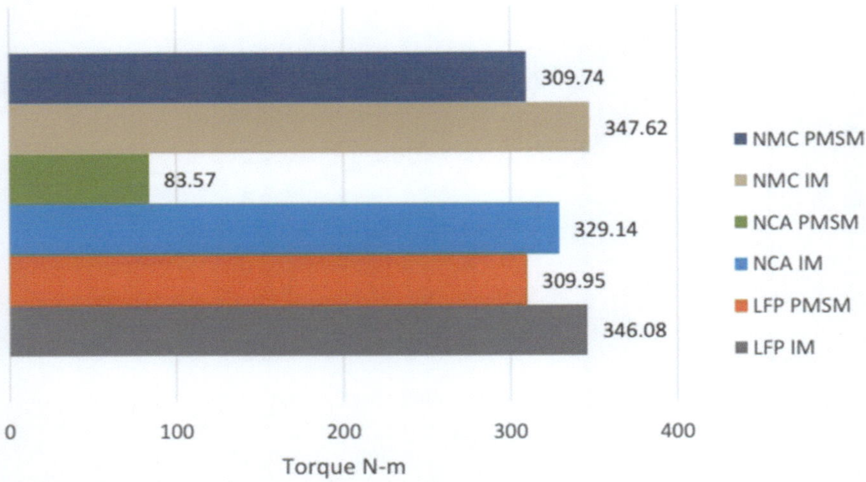

Fig. 1.3 Maximum torque for each combination

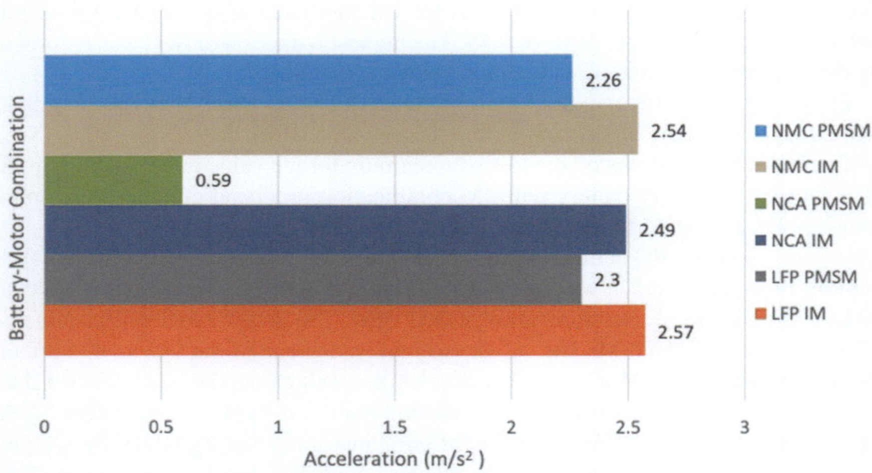

Fig. 1.4 Acceleration for all battery and motor combinations

The quicker the acceleration, the less time it took to attain a certain speed, resulting in a shorter distance covered during acceleration.

Table 1.3 shows the battery and motor combination comparison in terms of SoC, torque, acceleration and distance travelled from poor to best. Based on the table, IM performed very well compared to PMSM for most parameters. For the battery, the combination of NMC and IM showed good performance compared to the combination of LFP and IM. NCA on the other hand only showed good results for SoC but

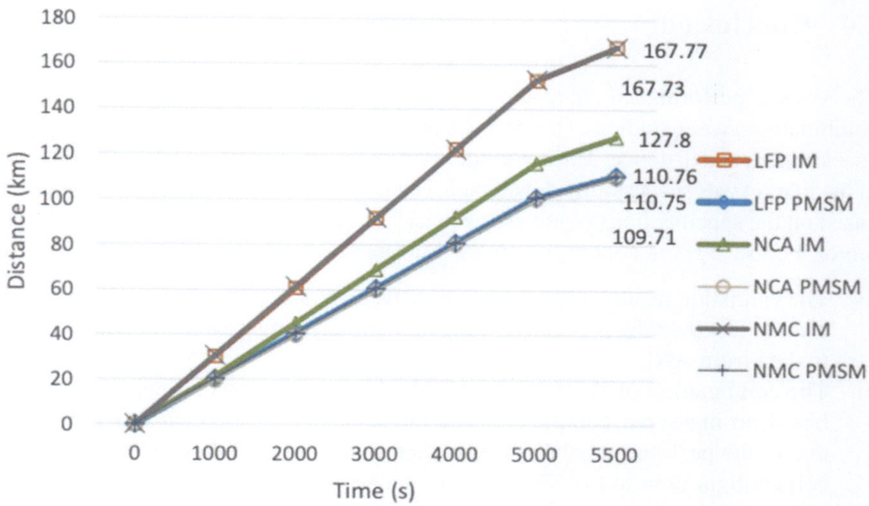

Fig. 1.5 Distance travelled by EV for each combination

was not so effective for other parameters making NMC the most optimal choice. But it should be noted that this simulation result was only for 5500 s of running time where it only affected 10–20% of the SoC.

Table 1.3 The most optimal to the poor combination for each parameter

Parameter	Motor-battery combination					
SoC	LFP IM	NMC IM	LFP PMSM	NMC PMSM	NCA IM	NCA PMSM
Torque	NCA PMSM	NMC PMSM	LFP PMSM	NCA IM	LFP IM	NMC IM
Acceleration	NCA PMSM	NMC PMSM	LFP PMSM	NCA IM	NMC IM	LFP IM
Distance	NCA PMSM	LFP PMSM	NMC PMSM	NCA IM	LFP IM	NMC IM

Poor → Best

1.4 Conclusion

The vehicle performance characteristics of EVs using different battery and motor combinations were studied. The performance parameters focussed on SoC, torque, acceleration, and distance. The EV model system based on existing EV specifications in the market was successfully developed. The EV model system had been developed based on the specifications of the 24 k Wh Li-Ion battery and also the 80 kW electric motor. Followings are conclusions made for this study.

i. The validation results was acceptable as the results obtained had low percentage error of less than 3% for each running mode when compared with the EV models to data from ANL.
ii. The combination of IM motor and NMC battery was the optimal combination based on measured parameters. The main factor that affected this result was due to the peak torque of IM was higher compared to PMSM. In addition, the cell configuration in the NMC battery pack contributed to power delivery also affected the obtained results.
iii. This research contributed to the growing body knowledge on batteries and motors in EVs. The findings highlighted the need to select appropriate battery chemistry and motor types to optimize EV performance and range. Further research into developing technologies and breakthroughs in battery and motor technology is recommended to drive the future of sustainable transportation.

References

1. S.H. Mohd Shafie, M. Mahmud, Urban air pollutant from motor vehicle emissions in Kuala Lumpur, Malaysia. Aerosol. Air. Qual. Res. **20**(12), 2793–2804 (2020)
2. Z. Chen, N. Guo, X. Li et al., Battery pack grouping and capacity improvement for electric vehicles based on a genetic algorithm. Energies **10**(4), 439 (2017)
3. A. Sakti, J.J. Michalek, E.R.H. Fuchs et al., A techno-economic analysis and optimization of Li-ion batteries for light-duty passenger vehicle electrification. J. Power. Sources **273**, 966–980 (2015)
4. S.J. Gerssen-Gondelach, A.P.C. Faaij, Performance of batteries for electric vehicles on short and longer term. J. Power. Sources **212**, 111–129 (2012)
5. A.D.W. Todd, P.P. Ferguson, M.D. Fleischauer et al., Tin-based materials as negative electrodes for Li-ion batteries: combinatorial approaches and mechanical methods. Int. J. Energy Res. **34**(6), 535–555 (2010)
6. M. Yildirim, M. Polat, H. Kurum, A survey on comparison of electric motor types and drives used for electric vehicles, in *Paper presented at the 16th International Power Electronics and Motion Control Conference and Exposition*, Antalya, Turkey (2014)
7. C. Iclodean, B. Varga, N. Burnete et al., Comparison of different battery types for electric vehicles. IOP. Conf. Ser-Mat. Sci. **252**, 012058 (2017)
8. G. Qiang, C. Xiusheng, Research on battery identification of electric vehicle battery management system, in *Paper presented at the International Conference on Computational and Information Sciences*, Chengdu, China (2010)

9. M.K. Tran, A. Dacosta, A. Mevawalla et al., Comparative study of equivalent circuit models performance in four common lithium-ion batteries: LFP, NMC, LMO, NCA. Batteries **7**(3), 51 (2021)
10. S. Lemon, A. Miller, Electric vehicles in New Zealand: technologically challenged? in *Paper presented at EEA Conference & Exhibition*, Auckland, New Zealand (2013)

Chapter 2
Enhancing Condition-Based Maintenance at a Malaysian Chemical Company Through the Use of Acoustic Emission Testing

Shahrizan Yusoff, Soleham Juraimi Kamaruddin, and Mohamad Sazali Said

Abstract The acoustic emission technique (AET) is widely used for detecting and monitoring damages and progress on damages in various structures, and it has acquired a reputation as one of the most dependable and well-established non-destructive testing techniques. Acoustic emission is a very efficient and effective technique for detecting fracture behaviour and fatigue in metals, fibreglass, wood, composites, ceramics, concrete, and polymers, among other materials. It can also be used to detect defects and pressure leaks in vessels, tanks, and pipes, as well as to track the progress of welding corrosion. A standard AET technique is also established in this study. The findings of this feasibility study can be utilised to support future offshore field tests of AET on operational structures. The findings revealed that the case company has a considerable need for systematic condition-based maintenance methods to do the right things at the right time and in a cost-effective manner.

Keywords Condition-based maintenance · Acoustic emission technique · Petro-chemical industry

2.1 Introduction

The petroleum refineries and chemical plants are large and complex operations with many systems and subsystems operating simultaneously. Within and in-between each system and subsystem are separators, heat exchangers such as condensers,

S. Yusoff (✉) · S. J. Kamaruddin · M. S. Said
Universiti Kuala Lumpur Malaysian Spanish Institute, Kulim, Kedah, Malaysia
e-mail: mshahrizan@unikl.edu.my

S. J. Kamaruddin
e-mail: soleham@s.unikl.edu.my

M. S. Said
e-mail: msazali@unikl.edu.my

T. A. Razak et al. (eds.), *IT Application for Sustainable Living 2*,
SpringerBriefs in Applied Sciences and Technology,
https://doi.org/10.1007/978-3-031-93198-7_2

boilers, and re-boilers, valves, scrubbers, accumulators, piping systems, and rotating mechanical systems. However, corrosion is a major concern in processing facilities. Each of these components has the potential to effect production, cause health and safety issues, and/or increase the environmental impact of the operations should failure occur [1].

Even though the processing plants are carefully designed, yet as ageing of equipment is becoming more and more a crucial issue. In the US for example, for instance, more than 50% of the pipeline system is over 40 years old [2]. Considerable quantity of underground in-plant pipelines of crude oil of fuel storage depots and refineries is operated for more than forty years. Corrosion damages and material degradation are the main causes for structural failures of such underground pipelines. Such failures not detected on time and not monitored in time are potentially the reasons of major breakdowns of the infrastructure of technological underground pipeline installations in-service and finally are the risks of pollution of the environment.

Acoustic emission testing (AET) has been used successfully to detect flaws in various structures during operation [3]. AET is used as a global, real-time monitoring, nondestructive testing (NDT) method for the assessment of the equipment integrity of large-scale structures with an emphasis on metallic pressure vessels, valves, cargo tanks and offshore structures.

This paper reports the application of AET to examine the viability of using acoustic emission technology as a screening tool for surveys and quantify the leak for valves in Malaysia. The main objective of applying this technology is to locate and identify the leak on different valves while a plant is operational. For that purposed, the acoustic emission (AE) source identification and techniques used are discussed. Then, a standard AET procedure, as (1) test plan, (2) AET system installation and checks, (3) data acquisition and analysis, (4) reporting documentation, and (5) follow-up inspection are established. The results from the implementation process can be used to plan improvement action further continuously and provide better maintenance activities in the future. Therefore, the overall optimal maintenance system can be achieved and drive the company to reach its mission and vision.

2.2 Literature Review

Establishing condition-based maintenance (CBM) strategies or methods for the repair or replacement of components at refinary plants can help minimise costly breakdowns. In agreement with the importance of asset management monitoring, the valve system's availability and reliability is a high level of the plant required to meet the production objective without interruption [4]. The valve is typically inspected before the hydraulic test, but not again until after the in-service time. This means that any problems detected during a new valve's hydraulic test may only be discovered after some time in service. While traditional NDT methods need the equipment to be taken out of operation and opened and cleaned for an inside examination, conventional NDT procedures do not. Jackets must be peeled, insulation or linings must be removed completely, and high-temperature vessels must be cooled to room temperature.

2.2.1 The Acoustic Emission Technique

Thus, advanced NDT testing such as the acoustic emission technique has the potential to overcome these constraints ensuring the continuous monitoring of sections of damaged pipeline during service without stop in transmission system. The AET is based on the phenomenon of radiation of acoustic (elastic) waves in solids that occurs when a material undergoes irreversible changes in its internal structure, for example as a result of crack formation or plastic deformation due to ageing, temperature gradients or external mechanical forces [5]. Moreover, AET monitoring also helps to prevent catastrophic failure of the valve leak that may prevent from loss of pressure containment.

According to [6] the vibration and dynamics noise from pressure-reducing devices usually generate acoustics with high energy capable of exciting vibration modes in pipe shells. Due to such high energy, the application of acoustic emission to non-destructive testing of materials typically takes place between 100 kHz and 1 MHz. Unlike conventional ultrasonic testing, AE tools are designed for monitoring acoustic emissions produced by the material during failure or stress, and not on the material's effect on externally generated waves. Part failure can be documented during unattended monitoring. In addition to non-destructive testing, acoustic emission monitoring has applications in process monitoring. Applications where acoustic emission monitoring has successfully been used include detecting anomalies in fluidized beds, and end points in batch granulation.

Figure 2.1 shows the basic principle of the AET. These structural alterations are the results of either internally generated or externally applied mechanical or thermal stresses. Numerous mechanisms exist for generation of AE, such as plastic deformation in materials, formation and growth of cracks, fibre breakage and delamination in composites, chemical reactions, phase transformations, impact, frictions, leakage, microseismic and seismic activities in geological materials.

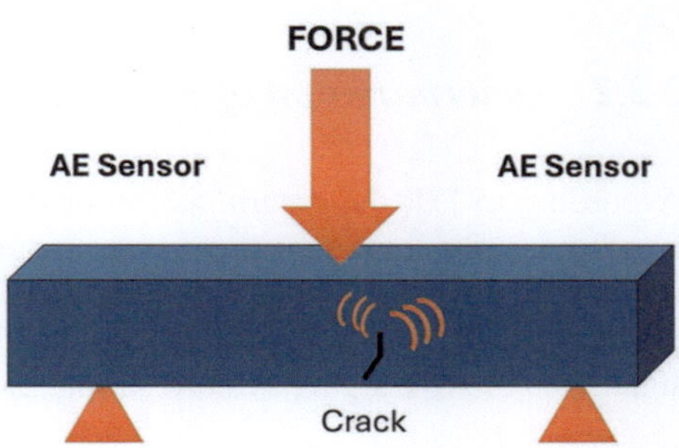

Fig. 2.1 Basic principle of the AET

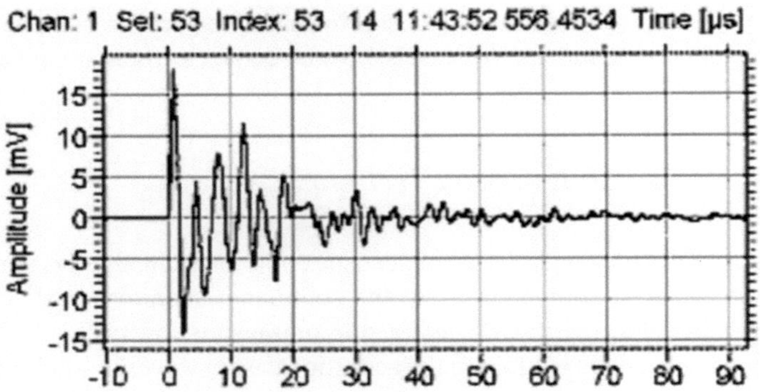

Fig. 2.2 Burst acoustic emission signal

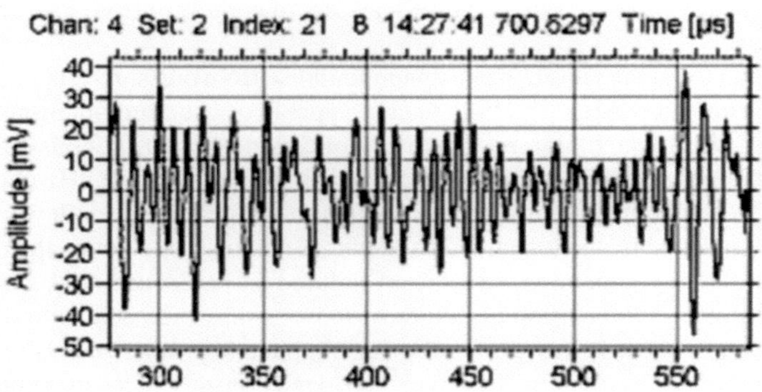

Fig. 2.3 Continuous acoustic emission signal

Most AET events are transient bursts of a few microseconds in duration, e.g. due to crack growth (see Fig. 2.2), but some appear to be 'continuous' and are mostly the superposition of large numbers of overlapping transient events, due to crack surface fretting as shown in Figs. 2.2 and 2.3.

2.2.2 Determination of the AET Signal

According to [7], the determination of the AET signal in the procedure of noise filtering is very important to separate indications from non-relevant indications or indications or signatures for structural defects. Moreover, it has been postulated for many years that different defects would leave unique wave form characteristic signatures of AET signal as show in Fig. 2.4.

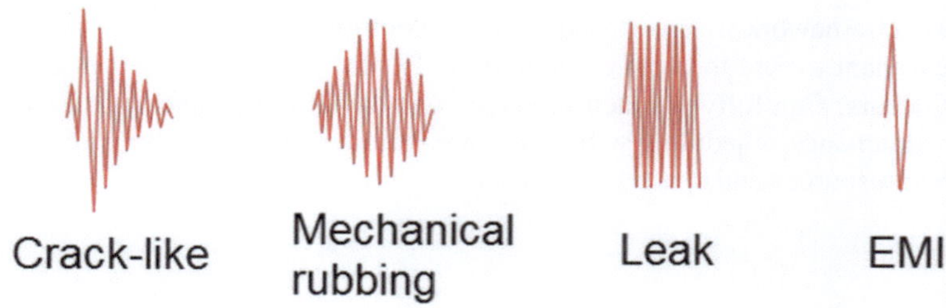

Fig. 2.4 Unique wave form (signature) for each defect

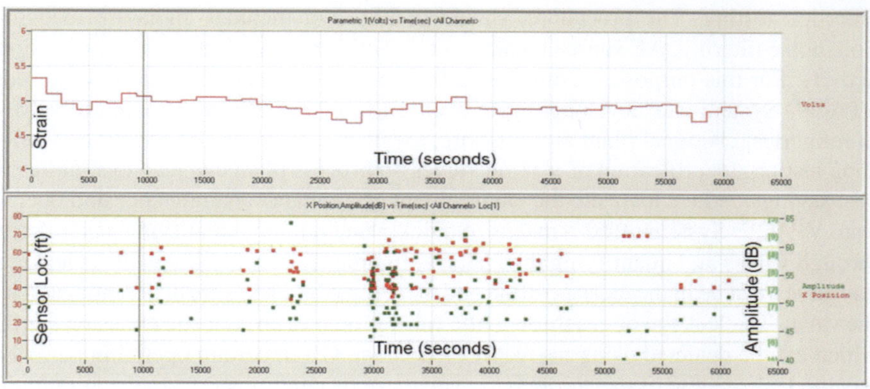

Fig. 2.5 Correlation plots for pattern recognition

In order to have the correct location of the defect, multiple, multiple sensors can be employs to determine where the acoustic emission signal originates from. The computed source locations are usually shown as points on a map-like display (*y* position vs. *x* position), with the sensor positions and/or structural features displayed in the background. Software is readily available for use on closed forms such as cylinders, cones, and spheres, as well as flat plates as shown in Fig. 2.5.

2.3 Methodology

In this study, the following factors are considered significant contributors in creating and maintaining the company's competitiveness:

1. High-quality product/service;
2. Delivery on time;
3. Environmental friendly production;
4. Society acceptance.

Thus, a framework was proposed to have an optimised maintenance system at the chemical plant to find the best combination of CBM and AET activities each time failure occurs. Carefully planned activities will avoid redundancy of tasks, while good maintenance scheduling will make sure that the company is prepared in terms of human resources and spare parts for the activities.

2.3.1 Analysis Procedure for Acoustic Emission Data

The acoustic emission data analysis procedure is critical to the success of the acoustic emission testing. The procedure, shown in Fig. 2.6, includes signature recognition, noise filtering, AE source location and identification of high acoustic emission activity. For this purpose, a computer software named as VPAC System was used. VPAC™ System was developed from over ten years experience listening to valves in refineries, chemical plant and offshore, together with extensive follow-up workshop testing. Identification of leaking valves on process plant can lead to significant cost savings, reduce losses to the environment and enhance maintenance and operation. VPAC™ System offers a means of quickly and reliably identifying gas leakage through valves and quantifying the leakage. In order to identify if a valve is leaking the valve must be in the close position and there must be a pressure differential across the valve. The leakage of gas through the valve produces an acoustic emission signal which can be detected using the VPAC™ System. Given certain operating parameters the leakage rate of gas through the valve can be determined from the amplitude of this signal.

In this case study there are a total of 18 different points for each set of valve measurements for AET. A device called call pocket AE was used in the field. This VPAC™ is a high performance, dual-channel acoustic emission system packed in a rugged, portable handheld unit optimal for use as a field survey tool. Moreover, the Pocket A VPAC™ offers all the performance, capabilities and features of a larger, more expensive AE system (wide bandwidth, speed, AE features, sampling rates, waveform processing, audible AE) in a compact, battery-operated package.

2.4 Result and Discussion

The results obtained from the acoustic emission technique for detecting valve passing during operation conditions can be found in Table 2.1.

Detail from site observations found that 8 out 9 inspected valves observed without passing issue. Only one valve, PDG321426 which is the 2nd Valve Inlet Economizer E3214 experienced passing about quite significant leakage rate at 1.4 L/min, which might cause losses to the MESH production (Fig. 2.7).

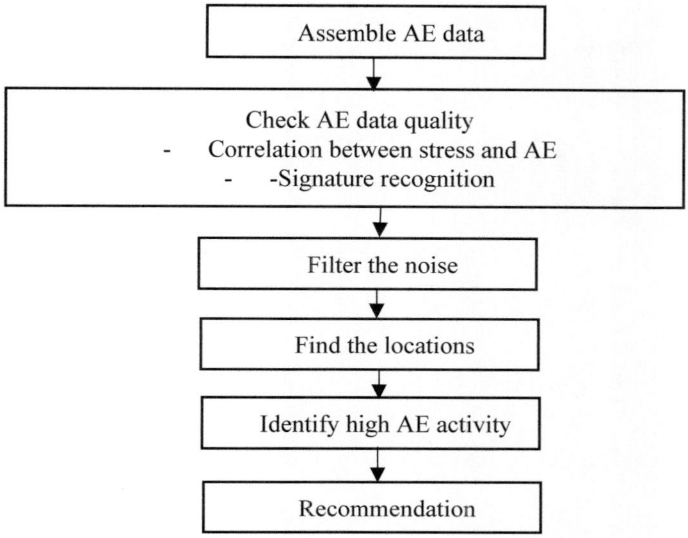

Fig. 2.6 Procedure of AET analysis

2.5 Conclusion

In conclusion, based on the AET assessment, this framework successfully ensures a continuous safe operation of the facilities and can prevent catastrophic failure and loss of containment. Some improvements can be implemented in the following study: installing a wireless condition monitoring sensor for monitoring valve failure online. In order to improve safety and reliability of the valve, a full scale comprehensive monitoring system by using the acoustic emission technique need to performed annually hence in can improve the plant reliability and availability.

Table 2.1 Detail of AET assessment result

Test point	Valve ID		Type of fluid	Up stream	Signal level (dB)	Down stream	Pressure Diff. (bar)	Inlet size (inch)	Type of valve	Leak rate (l/min)	Valce status	Mass density at standard condition (kg/m^3)	Loss (tonnes/year)
1	PT230002		Gas	12	12	12	27.8	0.75	Plug	–	Holding	–	–
2	Second manual valve vent to flare		Gas	12	12	12	27.8	1	Plug	–	Holding	–	–
3	PSV321006A		Gas	12	12	12	26.3	3	RV	–	Holding	–	–
4	PSV321006B		Gas	12	12	12	26.3	3	RV	–	Holding	–	–
5	XV321009		Gas	12	12	12	25.8	4	Ball	–	Holding	–	–
6	PT321015	First valve	Gas	12	12	12	26.0	0.75	Plug	–	Holding	–	–
		Second valve	Gas	12	12	12	26.0	0.75	Plug	–	Holding	–	–
7	PDG321426–Inlet E3214	First valve	Gas	12	12	12	26.0	0.75	Ball	–	Holding	–	–
		Second valve	Gas	17	41	36	26.0	0.75	Ball	1.4	Passing	1.36	1.0

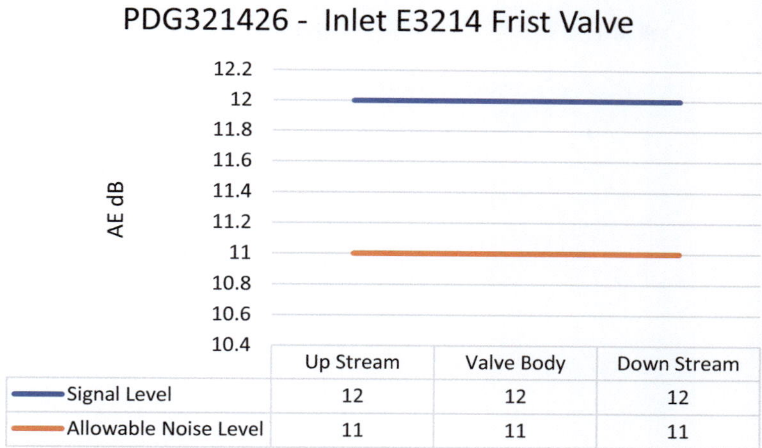

Fig. 2.7 AET signal level for the failure valve

References

1. R. Passler, Dispersion-related assessments of temperature dependences for the fundamental band gap of hexagonal GaN. J. Appl. Phys. **90**(8), 3956–3964 (2001)
2. J. Verran, Use of composite wraps to prevent acoustic-induced fatigue failure in piping systems, in *CEED Seminar Proceedings* (Curtin University, 2012)
3. M. Agar, L. Ancient, Acoustic-induced vibration: a new methodology for improved piping design practice, in *Offshore Technology Conference Asia*, Kuala Lumpur (2016)
4. M. Hamilton, Vibration-induced fatigue in piping, in *Spectrum Acoustic Consultants*, 28 April 2017 (Heat Transfer Handbook, USA, 2003)
5. The Piping Engineering. Guidelines for Acoustic Induced Vibration (AIV), Flow-Induced Vibration (FIV) analysis. http://www.piping-engineering.com/acoustic-induced-vibration-aiv-flowinduced-vibration-fiv-analysis-guidelines.html. Accessed 26 Jan. 2024
6. K. Kalantar-Zadeh, A. Kis, J.N. Coleman, M.S. Strano, Electronics and optoelectronics of two-dimensional transition metal dichalcogenides. Nat. Nanotechnol. **7**(11), 699–712 (2012)
7. A. Goswami, S. Kumar, Failure of water quench tower in an ethylene cracking plant. Eng. Fail. Anal. 104857 (2020)

Chapter 3
Development of a Smart Agriculture for Soil Moisture Measurement Level (Smart Farming)

Noor Iman Fitri bin Abu Bakar, Muhammad Afif Asnawi bin Muhammad Azizan, Muhammad Rahimy bin Mohd Radzi, Wan Nur Suhada Wan Mohd Zor, and Ahamad Zaki Mohamed Noor

Abstract This paper is focused on the implementation of a smart farming system. It should assist farmers to maintain the quality of farming products, farming production and farming market stability related to consumer. The concept of the project is to read the moisture condition inside the farming soil and to be able to interpret the data for condition monitoring. A high precision moisture sensor is placed at both equipment legs to penetrate the soil and to take the soil moisture reading. For this project, two methods were used to test the soil moisture, i.e., the Blynk App and ThingSpeak. The problem statement in this project is the connection between the product or circuit to devices or phone. By using the Blynk app and ThingSpeak, these methods showed that the sensor will detect that the soil is dry and transfer the information to the Blynk app and ThingSpeak for the users to see. The smart farming system will give an instruction to water the plant when it is in a red zone which means that the solid is in a dry state. A notification will be received by the users to tell them that they

N. I. F. bin Abu Bakar · M. A. A. bin Muhammad Azizan · M. R. bin Mohd Radzi · W. N. S. W. M. Zor
Mechanical Engineering Section, Malaysian Spanish Institute, Universiti Kuala Lumpur, Kulim, Kedah, Malaysia
e-mail: iman.abu23@s.unikl.edu.my

M. A. A. bin Muhammad Azizan
e-mail: afif.azizan25@s.unikl.edu.my

M. R. bin Mohd Radzi
e-mail: rahimy.radzi@s.unikl.edu.my

W. N. S. W. M. Zor
e-mail: suhada.zor@s.unikl.edu.my

A. Z. M. Noor (✉)
Manufacturing Section, Malaysian Spanish Institute, Universiti Kuala Lumpur, Kulim, Kedah, Malaysia
e-mail: ahamadzaki@unikl.edu.my

T. A. A. Razak et al. (eds.), *IT Application for Sustainable Living 2*,
SpringerBriefs in Applied Sciences and Technology,
https://doi.org/10.1007/978-3-031-93198-7_3

need to water their plant. If the users are not paying attention to the gauge before, the notification will attract their attention to look at the plant now. For the Blynk app, the Arduino serial monitor also will give an instruction to the consumer to stop watering when the soil is wet enough. ThingSpeak will record the data that is displayed in a graph. ThingSpeak and Blynk app are IoT analytics platform services that will give advantages in this project.

Keywords Smart farming · ThingSpeak · Blynk · IoT

3.1 Introduction

The management of the production process has altered because of technological advancements. The ability to outfit the agriculture industry with Internet of Things (IoT) technology has made it possible for them to link and communicate with one another to accelerate the process. Furthermore, when fully functional IoT-enabled industrial environments are established, a lot of potential could be realized [1]. As agriculture is one of the basics of life which provides the most crucial resources, the food industry is started by cultivating and providing jobs for humans. The food industry is also one of the pillars and strengths of the financial condition in a country's economy [2]. Thus, modern technology is involved in the agriculture industry as new tools for growing food in the most precise ways. The Internet of things (IoT) technology has become an evolution in the technology industry which gives a lot of potential benefits to the community. The basic idea of the IoT was to connect any device with physical substance to the Internet. Then, the Web of things (WoT) can connect sensors with the Web and translate it into useful data and information [3].

3.2 Literature Review

As refer to Dagar et al. [4] Arduino, which is being used in smart farming devices, is one of the most innovative technologies available today. For a long time, automated smart farming systems have been studied in the literature. This is due to the automated systems advantages of high throughput, high efficiency, and the lack of skilled manpower.

3.2.1 NodeMcu

The ESP8266 NodeMCU connected to Wi-Fi and talked with Thingspeak and Bylnk, two IoT platforms, using the HTTP protocol. Wi-Fi was used to process and transmit the environmental parameter data that was gathered from sensors to cloud and IoT platforms. Users were able to see the inside environment conditions in real time by using time-series visualization of environmental data on IoT systems. Furthermore, the Bylnk android application's control buttons can now be turned on or off remotely, allowing users to modify the way environment controllers function [5].

3.2.1.1 Soil Moisture Sensor

Many sensors are commercially available for sensing soil moisture. For example, watermark moisture sensors and electromagnetic soil moisture sensors are used for irrigation scheduling. MEMS moisture sensors and SHT15 humidity sensors are used for monitoring the moisture of concrete structures. To reveal the fine-grained, dynamic moisture changes in an outdoor landscape, a reactive soil moisture sensor network Although those sensors can measure the soil moisture, most of them are expensive which is not cost-effective to measure soil moisture of all plants in a greenhouse [6].

3.2.1.2 Water Valve

The solenoid valve and the valve body are the two main components of a solenoid valve. An electromagnetically induction solenoid has a coil around an iron base called the plunger. It may usually be open or closed at rest. A valve is usually opened, and a normally closed valve is closed in the de-energized state. The spiral is energized and produces a magnetic field as a current passes through the solenoid. The plunger induces a gravitational pull, moves it and overcomes the spring force.

3.2.2 Cloud Platform

ThingSpeak and Blynk apps will be used as monitor and data storage. These platforms help in accomplishing the smart hydroponic system. These are the platforms that are widely used for small project to be expanded to industrial usage [8].

3.2.2.1 Blynk

The study uses the Wi-Fi module, moisture sensor and Blynk application to monitor the moist condition of soil. The sensors are placed inside soil near the plant. Then the sensor turns the data into electric waves. Meanwhile, the Wi-Fi modem is used to send the data, and it will display it on the Blynk app that has been linked to the system. Thus, this proposed technique helps to inform the user that the plant need water, and it will help users to take the next actions [7]. Blynk is a mobile internet of things (IoT) platform that runs on both iOS and Android devices. It can interact over the internet with several microcontrollers, including the Node MCU ESP8266. The first section is the Blynk app, which can be used to keep an eye on a computer and display data on widgets [9].

3.2.2.2 ThingSpeak

ThingSpeak is an IoT analytics tool that helps to aggregate, analyze, and visualize data from connected devices. In the cloud, it can be used to simulate and interpret live data sources. One of the most critical elements of an IoT system is an IoT operation. ThingSpeak is an example of an app platform with a diverse set of capabilities [10].

3.3 Methodology

The flowchart of the smart farming project is shown in Fig. 3.1. For the equipment to generate moisture data reading, soil moisture sensor will be insert inside the soil by positioning the sensor leg accurately covering the soil area. For a more accurate reading, the sensor's entire leg was made sure to be inserted entirely into the ground. The earth is quite dry before the plant is watered. The Blynk app will display the information that the sensor detected that the soil is dry for customers to view. There is a gauge meter with a red reading of 100 in the Blynk app's UI. This suggests that there is dry soil. In this instance, a greater number on the gauge 100 indicates that the earth is extremely dry. Red denotes the soil and has aesthetic significance. After then, as soon as the plant is hydrated, the gauge's value will change. As of right moment, the gauge's value is 49 and is green.

The gauge's lower number and green color indicate to consumers that the soil is moist. Users will receive a second warning after 5 s of the modified value informing them to cease watering the plant if they have not already when the gauge meter reaches 45 or lower.

3.4 Result and Discussion

In this project mainly the physical testing was done based on a chili plant. This type of plant requires very close monitoring and cannot be abandoned even for an hour. By using this system, monitoring the chili plant condition will be easy. The Chili plant cannot have a very dry soil as the plant is so sensitive and may cause the plant to wither and die. However, the plant also cannot last with too much water as it will cause the chili to be wet and the moisture can cause disease toward the chili itself. Therefore, the coding we made in this product is when the gauge reaches 75, this is already very dry for a chili plant and the notification will be received via the Blynk app to water the plant.

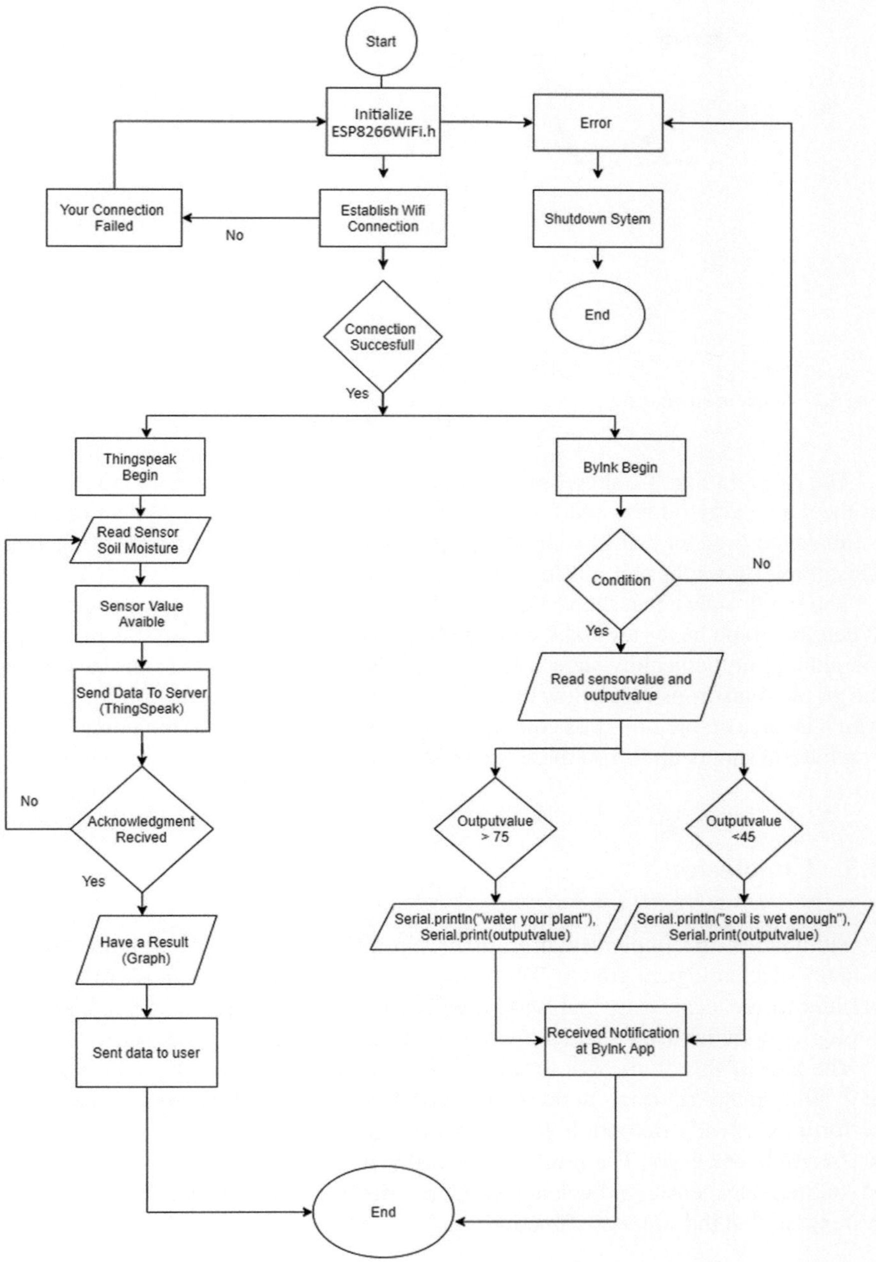

Fig. 3.1 Flowchart of smart farming

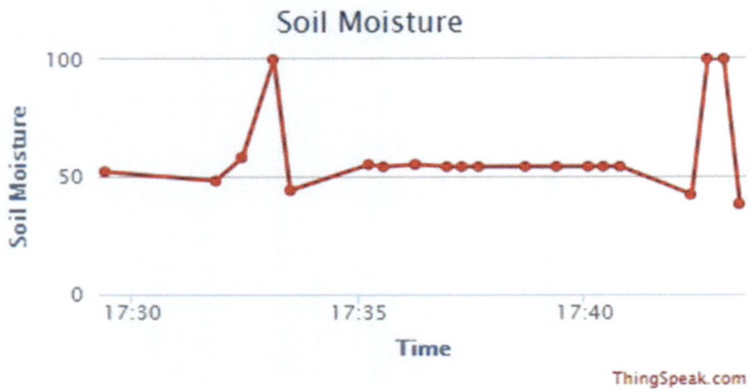

Fig. 3.2 Graph of soil moisture

The graph in Fig. 3.2 shows the soil moisture vs time. It will record the moisture of the plant every minute and form a graph. This can help users to collect data and estimate the time they must water the plant in future during certain time of the day. This graph shows the soil moisture around 5.30 pm to 5.40 pm.

The graph shows that the vertical axis is the soil moisture, i.e., larger when wet. When the graph hover around the middle which is near 50 as shown this mean that the plant is in the healthy state where the soil has its optimum moisture and when the graph trend drops way below this, it shows that the soil is overflowing with water which is undesirable or in bad condition. If the plant is kept dry to the point that the graph trend shoots up to 100 there is a chance that the plant will die or wither.

3.5 Conclusion

To summaries, the implementation of IoT in our everyday lives aims to boost a country's technological growth. This smart farming device is an example of a potential innovative technology that will be widely used. This project will also improve our agricultural industry's structure and mindset, especially in Malaysia.

The aim of this study was to use IoT technology to monitor the soil moisture of a chili plant. According to the findings of the study, the Blynk application could perform excellently on both IOS and Android devices. Basic widgets are available for free to Blynk users. The results suggested that the Blynk app can show the value of soil moisture sensor and when the button switches to ON a signal was sent to the systems so that the water are function.

References

1. M.H.F.M. Fauadi, A.A.M. Damanhuri, R.H. Hambali, A.Z. Mohamed, N.I.A. Noor, Recent applications of internet of things (IoT) in manufacturing sector–a review. Int. J. Sci. Technol. Res. **9**(09), 334–339 (2020)
2. C.A. Silva, D. Baker, A.W. Shepherd et al., Agro-industries for development, in *FAO Conference* (2009), pp. 96–97
3. P. Theekakul, S. Thiemjarus, E. Nantajeewarawat, T. Supnithi, K. Hirota, A rule-based approach to activity recognition, in *Lecture Notes Computer Science*, vol. 6746 (2010), pp. 204–215
4. R. Dagar et al., in *2018 International Conference on Inventive Research in Computing Applications (ICIRCA)* (2018)
5. Z. Wan et al., in 2019 *IEEE 3rd Information Technology, Networking, Electronic and Automation Control Conference (ITNEC)* (2019). https://doi.org/10.1109/ITNEC.2019.8729519
6. Wang et al., *Proceedings of the 18th International Conference on Mobile Systems, Applications, and Services* (2020)
7. Chang et al., *Proceedings of the ACM on Interactive, Mobile, Wearable and Ubiquitous Technologies* (2022); Serikul et al., in *2018 16th International Conference on ICT and Knowledge Engineering (ICT&KE)* (2018).
8. J.M.S. Waworundeng, N.C. Suseno, R.R.Y. Manaha, Automatic watering system for plants with IoT monitoring and notification. CogITo Smart J. **4**(2), 316 (2019)
9. Jamal et al., in *2020 59th Annual Conference of the Society of Instrument and Control Engineers of Japan (SICE)* (2020)
10. A.Z.M. Noor, M.S. Dzulkafle, N.M. Bohri, M.H.M. Khaireez, S.S. Jupri, M.H.F.M. Fauadi, Development of smart agriculture (smart hydroponic) to monitor soil humidity level, in *IT Applications for Sustainable Living. SpringerBriefs in Applied Sciences and Technology*, ed. by M.H. Abu Bakar, T.A. Abdul Razak, A. Öchsner (Springer, Cham, 2023) pp. 25–34

Chapter 4
Development of a Driver Drowsiness Alert System Under Extreme Lighting

Alya Nabilah Mohd Nodri, Mohd Fauzi Abu Hassan, Zakiah Ahmad, Jum'azulhisham Abdul Shukor, and Mazwin Tan

Abstract Drowsiness is one of the many leading causes of death and injury in tragic road accidents all around the world. However, its early indicators such as weariness, fatigue, or microsleep should be noticed earlier before any catastrophic accidents occur. Thus, a driver drowsiness alert system was developed to detect driver drowsiness. However, detecting drowsiness in extreme lighting situations may develop a problem as the image captured may generate much noise thus, delaying the alert reaction time to the driver. This paper will initiate an upturn on those drawbacks with the aid of adaptive gamma reduction with weighting distribution.

Keywords Drowsiness · Pre-processing · Adaptive gamma reduction · Extreme lighting

---

A. N. M. Nodri · M. F. A. Hassan (✉)
Intelligent Automotive Systems Research Cluster, Electrical, Electronics and Automation Section, Universiti Kuala Lumpur Malaysian Spanish Institute, Kulim, Kedah, Malaysia
e-mail: mohdfauzi@unikl.edu.my

A. N. M. Nodri
e-mail: alya.nodri25@s.unikl.edu.my

Z. Ahmad
Engineering Section, Universiti Kuala Lumpur Malaysian Spanish Institute, Kulim, Kedah, Malaysia
e-mail: zakiah@unikl.edu.my

J. A. Shukor
Manufacturing Section, Universiti Kuala Lumpur Malaysian Spanish Institute, Kulim, Kedah, Malaysia
e-mail: azulhisham@unikl.edu.my

M. Tan
Mechanical Section, Universiti Kuala Lumpur Malaysian Spanish Institute, Kulim, Kedah, Malaysia
e-mail: mazwin@unikl.edu.my

T. A. A. Razak et al. (eds.), *IT Application for Sustainable Living 2*,
SpringerBriefs in Applied Sciences and Technology,
https://doi.org/10.1007/978-3-031-93198-7_4

4.1 Introduction

The number of automobiles in the nation is rising rapidly throughout the year. The growth in numbers relates to the increasing number of traffic accidents with the increased traffic. Until now, numerous studies have been conducted to address the risk of safety posed by drowsy driving, which can be divided into five categories: subjective, behavioral, physiological, vehicle-based, and hybrid evaluation [1]. The discovery of a hybrid assessment technique that includes the use of image processing has been a great contribution to the development of the advanced driving assistance system. Even though it has undeniably enhanced the system's detection accuracy, consistency, and speed, the studies by Suhaiman et al. [8] have discovered several limits which affect the accuracy of the camera's ability to measure the driver's face and eye. That is due to the high variability of surrounding parameters. A camera that works under extreme lighting conditions may generate much noise that will delay the reaction time of the system to give an alert to the driver in real-time as of the weak resolution and contrast degradation from the captured image [8–10].

Therefore, when it comes to accurately detecting driver fatigue using visual features, the intelligent system often encounters difficulties. This paper will begin addressing the issues by creating a module that works on enhancing the captured image. This will be accomplished through image pre-processing techniques to make it simpler for the system to analyze the driver's state from the captured image. The remaining part of this paper is organized as follows: Sect. 4.2 examines significant reviews of the previous research field. While the module's structure and test procedure will be discussed more in Sect. 4.3. The results of the test that were done are shown in Sect. 4.4. Lastly, the conclusion and future work were drawn in Sect. 4.5.

4.2 Literature

The critical issue that a drowsiness system must address is the question of how accurate detection of drowsiness can be done at an early stage. That is why image quality is a crucial consideration in image-based object recognition, pre-processing is the key missing element in the system's algorithm. Figure 4.1 shows the flowchart of the three main stages of the pre-processing section of the algorithm. Despite its importance or usefulness when each of these algorithm's used alone, all three of the steps in the pre-processing stages play the same importance in producing the best method for more effective image processing. In fact, it is necessary to implement a bit of every method to the steps involved. The enhancement at the third stage will not increase the original information content of the data, instead, it will increase the dynamic range of specific features as chosen [4].

The existing contrast enhancement techniques can be categorized into three groups: global, local, and hybrid techniques. Digital images' quality and appearance have been improved through efforts to improve image resolution (visual appearance).

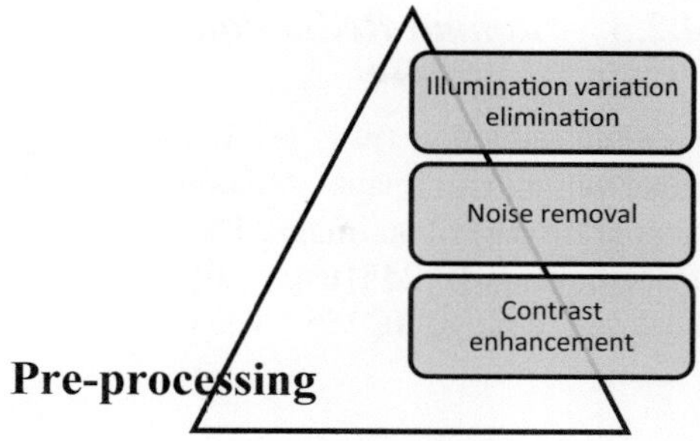

Fig. 4.1 Pre-processing stages in an algorithm

During the process of quantifying the entire work, the idea that the number of pixels is the result of sampling as well as the effect of gray level or intensity was taken into consideration [5].

4.2.1 Brightness and Contrast Enhancement

Brightness is the preset level of luminance in an image while contrast is defined as the difference in brightness (degree of intensity) between an image of dark and light areas. In global enhancement techniques, a single adjustment function is applied to each output pixel value of an image [7]. The adaptation of global approaches may result in over or under-enhancement problems for certain parts of an image since each part of the image requires a specific definition of enhancement. Contrast enhancement is an important area in image processing where it is widely used for medical image processing and as a pre-processing step in speech recognition, texture synthesis, and many other image/video processing applications.

4.2.2 Histogram Equalization

Histogram equalization (HE) is one of the pixel brightness transformation techniques. By stretching the image's intensity range or dispersing the most frequent pixel intensity values, it offers a thorough way for image contrast [6]. It is so famous due to its effectiveness on practically almost all types of images. But sometimes that is not the case as it cannot adapt to the pixel intensity value of certain parts which requires a different approach [9].

4.2.3 Gamma Reduction

Gamma reduction (γ) is responsible for performing a non-linear method to every pixel value in an image. Although it only happens in background; it does affect the histogram data of an image. The value of pixel from an image will first be converted from the range [0, 255] to [0, 1.0]. Then after calculations, the values will be converted back to the range [0, 255]. The formulation can be referred as below:

$$O = \left(\frac{I}{255}\right)^{\frac{1}{\gamma}}.255 \tag{4.1}$$

where I denotes the input pixel value by [0, 255]. Then, the output pixel value, O of the input image which also in the same range [0, 255]. While gamma, γ that controls the intensity of light in the image pixels. If the gamma < 1, then the image will be darker and if the gamma value is > 1, the image will be lighter. That makes an image with a gamma value = 1 will gain no effect [10]. Adaptive gamma reduction (AGR) as stated before, the main drawback of the traditional gamma reduction method is to manually give the gamma value for image enhancement which is time-consuming [4] then solved the problem by implementing the weighted distribution method to the technique. Together with general histogram modification, the gamma reduction technique's new transformation makes a new formulation:

$$T(l) = I_{\max}\left(\frac{l}{I_{\max}}\right)^{\gamma} = I_{\max}\left(\frac{l}{I_{\max}}\right)^{1-\mathrm{cdf}}, \tag{4.2}$$

where $I_{\max}$ denotes the highest input value density. After applying Eq. (4.2), the intensity, I of each pixel in the input image is transformed to $T(l)$. Figure 4.2 shows the flow chart of the new gamma reduction transform.

4.3 Methodology

The main objective of developing the algorithm is to process images under extreme lighting conditions. Thus, the module should be able to differentiate between bright images and dark images. Then, the dark image will undergo cumulative distribution function (CDF) of their greyscale intensities together with the adaptive gamma reduction. An image is directly captured by the camera in the earlier driver drowsiness alert system. This newly developed module would start from that point onwards. A sample image exposed to unstructured lighting will be used in building the module for pre-processing.

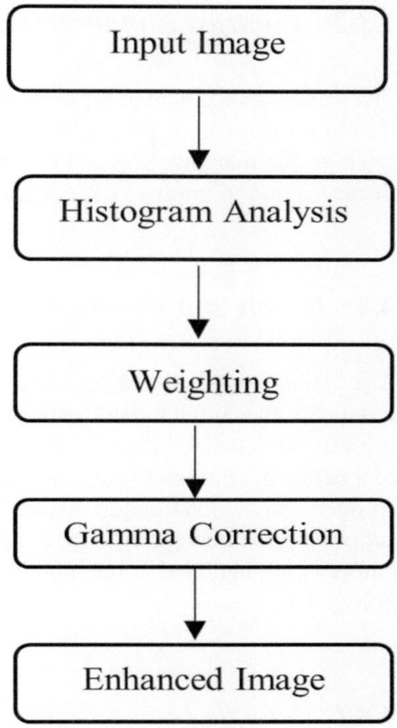

Fig. 4.2 Flowchart of gamma reduction with weighted distribution

4.3.1 Image Preparation

Before the sample image is utilized further, it will be prepared first by going through some steps which will be discussed further in the next part. Gray scale transformation in preparing the image before contrast enhancement starts as it aids by streamlining methods and removing issues associated with computational demands. A RGB image generally requires 24 bits for each pixel. Each channel for red, green, and blue requires 8 bits. To put it another way, a RGB image is made up of a total of three mages at one time. However, since a grayscale image consists of one image, the system only needs 8 bits for each pixel. A crucial pre-processing step before beginning a contrast enhancement is resizing the images. By using one standard size for all images, processing becomes quicker and requires less time. In this module, all images were set to 500 × 500 scale.

4.3.2 *Contrast Enhancement*

The word enhancement refers to increase or to improve which describes anything that is an improvement to the quality or value of something. This contrast enhancement refers to the manipulation and redistributing of the image pixels in a linear or non-linear fashion to improve the contrast of an image.

4.4 Result and Discussion

After conducting a literature survey, a suitable image enhancement technique was found. It is then implemented in the module for the pre-processing stage. A qualitative test was also conducted to test the performance of the module with a setup consisting of a personal computer (with an Intel® Core™ i5-8250U and 16 GB of RAM) and an open source code editor Visual Studio Code (famously known as VS Code) for supporting the debugging. The module was tested first with several sample images before being included in the module created. Its process can be described as shown in Fig. 4.3.

4.4.1 *Sample Image Before and After Pre-processing*

Figure 4.4 reveals that none of the sample input images show anything accept the darkness or total black while the sample output not only succeeds in improving the contrast but also succeeds in capturing the eyes and mouth which are the most vital facial feature for the driver drowsiness alert system to recognize. Figure 4.5 shows sample output images after pre-processing.

4.4.2 *Qualitative Analysis Using Histogram-CDF Graph*

After capturing both sample input and output images, a quantitative analysis is done to test the developed module's quality. Output graphs have a greater spread compared to their input graphs respectively (see Fig. 4.6).

The histogram part of the graphs shows the output image spans the full range of gray-level values 40 while the input images only manage to reach the value of 10 at most. Figure 4.7 shows sample output histogram-CDF graphs.

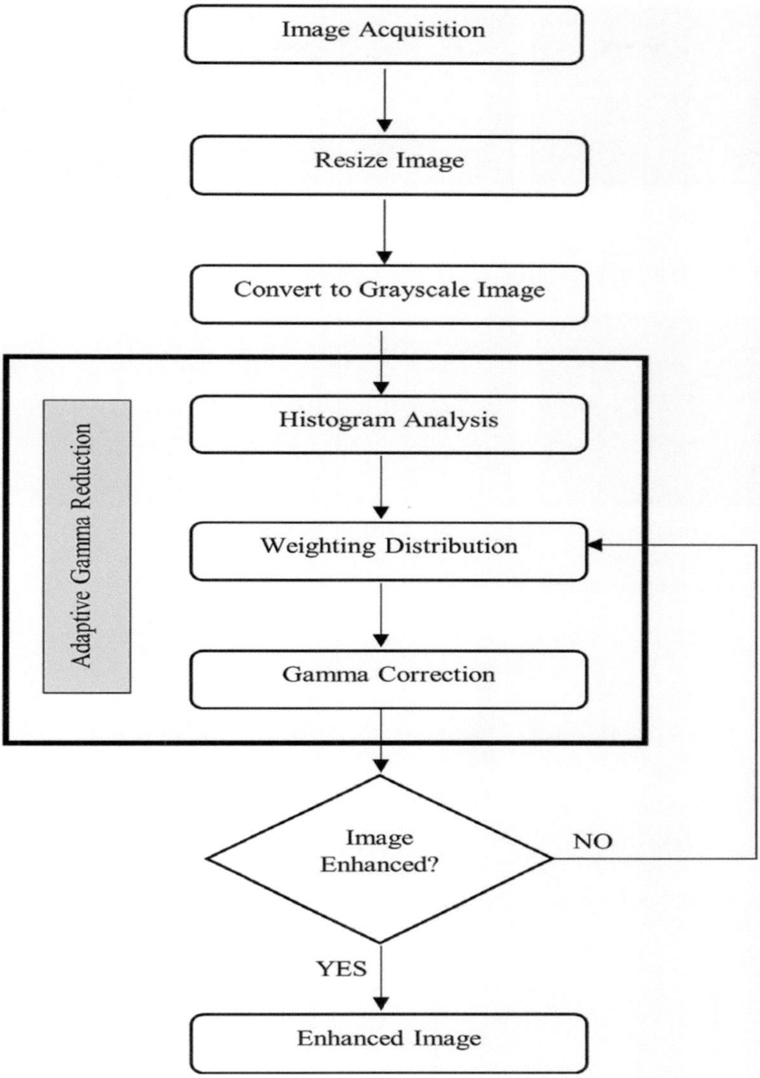

Fig. 4.3 Module flowchart

4.5 Conclusion

The adjustment to the pre-processing method using the adaptive gamma reduction method to the proposed module shows great improvement in the quality of output images. As the objectives of this paper have been successfully achieved, it can be said that this module is practical and applicable to be included in the previous driver

(a) (b) (c)

Fig. 4.4 Sample input images

(a) (b) (c)

Fig. 4.5 Sample output images

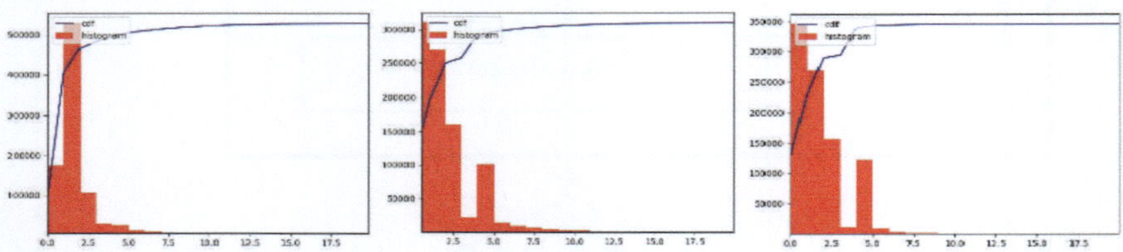

(a) (b) (c)

Fig. 4.6 Sample input histogram-CDF graph

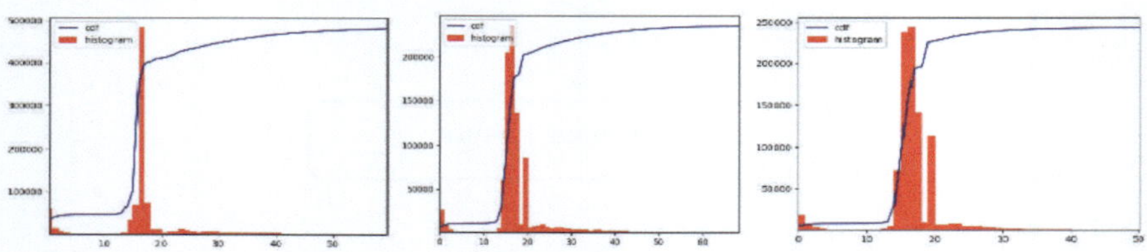

(a) (b) (c)

Fig. 4.7 Sample output histogram-CDF graphs

drowsiness alert system. This approach, however, is only applicable to images that are exposed under unstructured lighting. Other issues with acquiring the driver's visual features include images that are obscured or have low quality since the built-in capabilities of the camera have been left due to a lack of time and resources. Future development could involve various approaches as well as a deeper understanding of the system.

References

1. T. Azim, M.A. Jaffar, A.M. Mirza, Fully automated real time fatigue detection of drivers through fuzzy expert systems. Appl. Soft Comput. J. **18**, 25–38 (2014). https://doi.org/10.1016/j.asoc.2014.01.020
2. G. Cao, L. Huang, H. Tian, X. Huang, Y. Wang, R. Zhi, Contrast enhancement of brightness-distorted images by improved adaptive gamma correction (2018)
3. P. Chandra, S. Kumar, G.P. Sunkara, in *Advanced Image Processing Using Histogram Equalization and Android Application Implementation* (2016)
4. M. Dailla, P. Kaur, V. Dhawan, Adaptive gamma correction with weighted distribution and recursively separated and weighted histogram equalization: a comparative study (2014). www.ijert.org
5. R. Maini, H. Aggarwal, in *A Comprehensive Review of Image Enhancement Techniques*, vol. 2 (2010)
6. B.A. Nguchu, Critical analysis of image enhancement techniques. Int. J. Electr. Electron. Res. **4** (2016). www.researchpublish.com
7. S. Rahman, M.M. Rahman, M. Abdullah-Al-Wadud, G.D. Al-Quaderi, M. Shoyaib, An adaptive gamma correction for image enhancement. Eurasip J. Image Video Process. (2016)
8. A.A. Suhaiman, Z. May, N.A.I.A. Rahman, Development of an intelligent drowsiness detection system for drivers using image processing technique, in *2020 IEEE Student Conference on Research and Development*, SCOReD 2020 (2020), pp 233–236. https://doi.org/10.1109/SCOReD50371.2020.9250948
9. A. Toet, T. Wu, Efficient contrast enhancement through log-power histogram modification. J. Electron. Imaging **23**(6), 063017 (2014). https://doi.org/10.1117/1.jei.23.6.063017
10. M. Yaakop, S. Yaacob, A.H.F.S.A. Jamil, S.A. Bakar, M.F.A. Hassan, A.S. Pri, Detection of drowsiness using EEG probes sensory logic signals activeness topology, in *3rd IEEE International Conference on Artificial Intelligence in Engineering and Technology, IICAIET 2021* (2021). https://doi.org/10.1109/IICAIET51634.2021.9573986

Chapter 5
Development of Wireless Pulse Monitoring System for Hospital Patients

Mohd Fauzi Alias, Ahmad Syazili Abdul Rahim, Mohamad Rosyidi Ahmad, Mohd Suyerdi Omar, and Mohd Riduan Ibrahim

Abstract This study presents the design and development process of a novel integrated device designed for the purpose of monitoring the heart rate. The high cost of the pulse monitor may be attributed to the intricate nature of its system, hence limiting its widespread use mostly to healthcare facilities such as hospitals and clinics. The objective of this project is to design a bracelet that may be used in medical applications, facilitating user-friendly health monitoring. This will be achieved by using an Arduino Nano microcontroller, a photosensor, a Wi-Fi module, and a 3D printed shell. In addition to this, the objective of this project is to design and fabricate a dependable apparatus capable of yielding precise measurements for the assessment of heart rate using cost-effective photosensors. This project also includes the development of a graphical user interface intended for healthcare professionals, such as nurses or doctors, to effectively monitor the patient's heart rate expressed in beats per minute. The performance of our device was compared with that of an oscillometer device via the measurement of 10 individuals. The findings indicated that the device's error rate is insignificant and proves that the device is quite accurate and acceptable in measuring or monitoring the heart rate continuously.

Keywords UI · 3D printing · Photosensor · Oscillometer · BPM

M. F. Alias (✉) · A. S. A. Rahim · M. R. Ahmad
Electrical, Electronic and Automation Section, Universiti Kuala Lumpur, Malaysian Spanish Institute, Kulim, Kedah, Malaysia
e-mail: fauzialias@unikl.edu.my

M. R. Ahmad
e-mail: mrosyidi@unikl.edu.my

M. S. Omar
Mechanical Section, Universiti Kuala Lumpur, Malaysian Spanish Institute, Kulim, Kedah, Malaysia
e-mail: msuyerdi@unikl.edu.my

M. R. Ibrahim
Manufacturing Section, Universiti Kuala Lumpur, Malaysian Spanish Institute, Kulim, Kedah, Malaysia
e-mail: mohdriduan@unikl.edu.my

T. A. A. Razak et al. (eds.), *IT Application for Sustainable Living 2*, SpringerBriefs in Applied Sciences and Technology,
https://doi.org/10.1007/978-3-031-93198-7_5

5.1 Introduction

The heart rate serves as an indicator of cardiac health and aids in the evaluation of the cardiovascular system's state [1]. The heart rate is often assessed in a clinical setting using controlled methods such as blood measurement, heart auscultation, and electrocardiogram (ECG) [2]. One area of growing significance for wearable computers is their role as sophisticated medical monitoring devices, which provide real-time input to patients. The system has significant promise due to its wide applicability in the fields of medical, sports, and security. Due to the current rise in population and the consequent prevalence of chronic illnesses resulting from lifestyle modifications, there exists a pressing need to regularly assess the health condition of people in their day-to-day lives as a preventive measure against life-threatening ailments [3]. The Holter and oximetry machine are often used in medical settings, such as hospitals and clinics, for the purpose of monitoring pulse rate, oxygen saturation levels, and blood pressure. The objective of this study is to evaluate the optimal photo-sensor, including light emitting diode (LED), infrared (IR), and light dependent resistor (LDR), for detecting significant heart pulse signals from various parts of the human body that possess a high density of blood capillaries.

The wristband is equipped with a photoplethysmography (PPG) sensor and a microprocessor, enabling it to accurately detect the patient's pulse in beats per minute (BPM). The system is responsible for processing the data obtained from the sensor and then transmitting the heart rate information over a Wi-Fi connection. The system has four fundamental components namely a PPG sensor, a CPU, a Wi-Fi module, and three 3 V lithium-ion coin cell batteries. The capacity to use a limited number of components enables its complete integration into a 56.8 mm $\times$ 28 mm $\times$ 32 mm enclosure fabricated by 3D printing. The graphical user interface (UI) that we developed will primarily display the beats per minute of the patient who is wearing the device.

5.2 Literature Review

5.2.1 *Pulse Detection*

Prior research has been conducted on the photo sensor intended for use in the wristband, as well as investigates the absence of a previous model for the pulse detection and monitoring system. The pulse detector in heart rate monitors has two components, the pulse detecting unit and the heart rate displaying unit. The device used LEDs and a photosensor to quantify an individual's heart rate by detecting variations in blood reflectance [4]. The power emitted by the LEDs is adjusted to align with the characteristics of the photo sensor, resulting in a fluctuation in resistance within the detectable range of the photo sensor after attenuations caused by the presence of the index finger. The emitted light from the LED undergoes diffuse scattering and passes through the tissue of the fingertip. A light dependent resistor (LDR) or photosensor is placed on the skin's surface to measure the light that passes through the skin. The blood has a high absorption capacity for light, while tissue shows a comparatively lower absorption capacity [2]. The photodetector observes changes in light absorption resulting from fluctuations in blood volume flowing through the arteries as the light traverses the skin. This methodology enables the computation of the length of the cardiac cycle, often referred to as heart rate, by using the peaks of light absorption [3]. The PPG sensor employs a green LED with a wavelength of 565 nm and a photodetector that exhibits a heightened sensitivity to this particular wavelength. The selection of a green light is based on its predominant absorption characteristics that align with the red blood cells.

5.2.2 Pulse Monitoring System

The Holter electrocardiogram (ECG), developed by Norman Holter has significantly advanced the field of arrhythmia diagnosis. Arrhythmias refers to pathological conditions characterised by irregularities in the normal rhythmic cardiac contractions. The data for the Holter ECG is acquired using a monitor, which is used to record ECG readings continuously for a duration of 24 h. In practical application, it presents a challenge for a human observer to thoroughly analyse the whole long-term ECG and then arrive at a judgement [5]. The monitors are characterised by their substantial size, with each electrode being linked to the central unit via cables as shown in Fig. 5.1. Hence, the use of the monitor disrupts the patient's everyday regimen and is impractical for inconspicuous and uninterrupted monitoring.

Oscillometric measuring devices use an electronic pressure sensor that provides a numerical representation of blood pressure. Typically, the inflation and deflation of the cuff are facilitated by an electrically powered pump and valve, which may be positioned either at the wrist, raised to the level of the heart, or alternatively, at the upper arm, with a preference for the latter. The cuff is first inflated to a pressure that exceeds the systolic arterial pressure, after which the pressure is subsequently reduced to a level below the diastolic pressure. When blood flow is present but limited, the pressure exerted by the cuff will regularly fluctuate in accordance with the rhythmic expansion and contraction of the brachial artery. The systolic and diastolic pressure readings are derived from the raw data via an algorithm. However, this device has a substantial size and poses challenges in terms of its suitability for continuous real-time monitoring over a 24-h period.

Additional techniques for measuring heart rate and blood oxygen levels in the human body include the use of a pulse oximeter. This device uses an optical sensor and employs LEDs to produce light at varying wavelengths. The LED light will be emitted via a finger, and afterwards, the transmitted light will be detected by using an optical sensor. The determination of oxygen saturation level relies on the idea of differential light absorption by oxygenated and deoxygenated haemoglobin.

Fig. 5.1 Holter machine

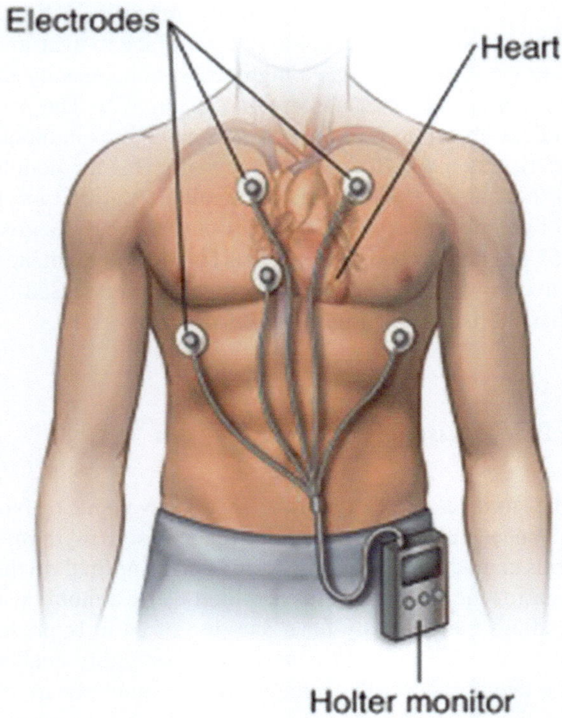

Oxygenated haemoglobin has a larger absorption coefficient for infrared light, whereas deoxygenated haemoglobin tends to absorb more red light (Fig. 5.2).

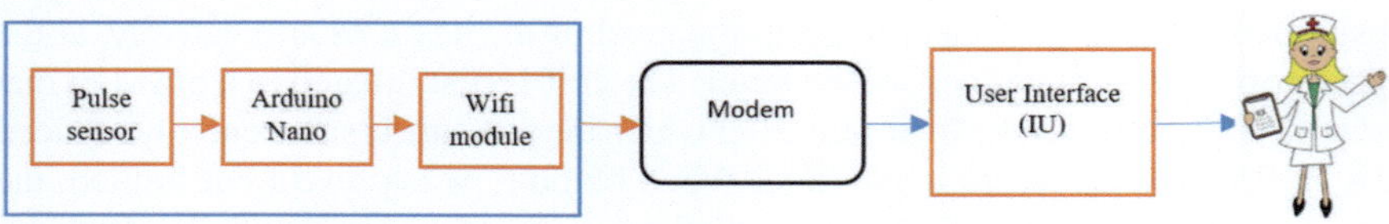

Fig. 5.2 System flowchart

5.3 Methodology/Experimental Set-Up

5.3.1 *Working Operation*

There are several types and specifications of the components that have been used for the wristband monitoring system in the market such as a wristband that uses the Arduino Nano microcontroller, a reflective pulse sensor, and a Wi-Fi module for the purpose of transmitting data to a computer. In the first stage, the pulse sensor will be responsible for detecting the pulse. The technique used for pulse detection is reflectance. The photosensor will detect the signal and then transmit the analogue data to the microcontroller, namely the Arduino Nano. The microcontroller is responsible for converting the analogue value into a corresponding value in beats per minute. Subsequently, the wireless module will transmit the signal to the computer through Wi-Fi connection. Once the signal has been sent to the computer, the graphic user interface (UI) will display the BPM graph. The primary objective of the UI is to provide continuous and real-time patient monitoring for healthcare professionals, namely doctors and nurses. The Arduino Nano microcontroller is shown in Fig. 5.3.

Based on the system's flowchart as shown in Fig. 5.2, pulse sensor generates an analogue signal that spans from zero to five volts, which is contingent upon the level of light detected by the photodetector. The incoming signal is subjected to processing by the CPU, wherein an adaptive peak detection algorithm is executed. The technique used in this study aims to standardise the signal within a range of 0–512, while simultaneously establishing the peak detection threshold at a value of 256. Once the threshold is surpassed for the first occurrence, the real peak value is recorded, subsequently followed by the matching trough. The newly established threshold is determined as the point equidistant from both the highest and lowest points. After the first detection of the first beat, a timer will commence and continue for a maximum duration of two seconds. If a peak is not identified within the specified time range, the

Fig. 5.3 Arduino nano microcontroller

subsequent beat will be regarded as the first one. Alternatively, the temporal interval between successive beats is measured upon the detection of a second beat.

5.3.2 Hardware Configuration

The wristband device uses a limited number of hardware components to avoid the risk of wire entanglement and ensure that the device is sufficiently tiny for patients to wear on their wrists. The software is designed to facilitate the compilation and transfer of source code to an Arduino Nano microcontroller, similar to the functionality provided by the Arduino Integrated Development Environment (IDE). The user interface component will be implemented using the Java Processing Integrated Development Environment (IDE). The use of the smallest available Arduino microcontroller on the market facilitates the conservation of device area, hence enabling the incorporation of additional components. The Arduino Nano is also capable of interfacing with analogue input and UART devices, such as the ESP8266.

The ESP8266 is compact in dimension and consumes little power. The ESP8266-05 requires minimal external circuitry for operation and incorporates a 32-bit Tensilica MCU, power amplifier, low noise receive amplifier, filters, and power management modules in a compact package. This Wi-Fi module can connect to the external antenna to improve the signal strength. The power-saving architecture includes three modes of operation: active mode, slumber mode, and deep sleep mode, allowing battery-powered designs to operate for extended periods of time.

The housing for the bracelet is fabricated using a 3D printer, as seen in Fig. 5.4. The inflexible housing is composed of acrylonitrile butadiene styrene (ABS) plastic, manufactured by the Makerbot brand. In the process of designing, each individual component is meticulously drafted using the Catia software. A meticulous approach is necessary to ensure the seamless integration of internal electronics with the designed elements. It is recommended to save the Catia pieces in.STL file format and thereafter import them into the Makerware software for the purpose of 3D printing. A certain degree of tolerance is often allowed for misprints, and the ultimate design often requires a process of experimentation and refinement. The ultimate configuration used for this product has a box structure, wherein the bottom component is affixed by the utilisation of ABS slurry. ABS slurry is a composite material composed of more ABS plastic dissolved in an equivalent proportion of acetone. In order to produce ABS items, it is necessary to use a heated build plate set at a temperature of 110°C, which should be covered with Kapton tape. Additionally, the temperature of the extruder nozzle should be adjusted to around 225 °C

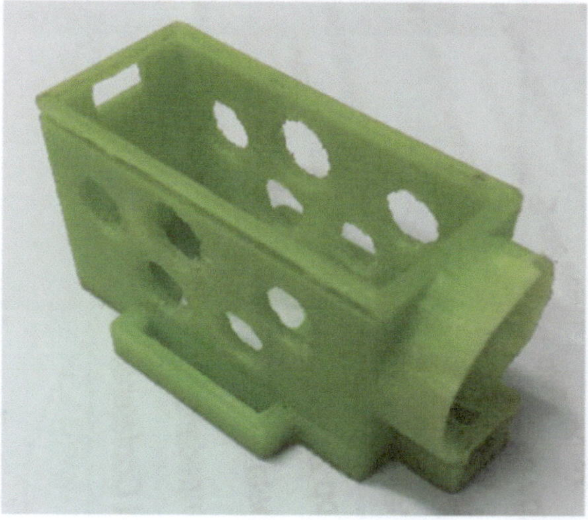

Fig. 5.4 Microcontroller housing 3D printed model

5.3.3 *Software Configuration*

The processing IDE is being used for the graphical user interface of the project. The programming language used is C++, which has similarities with the Arduino programming language. The system visually represents the user's pulse in real-time via a graphical display. Additionally, it generates a graphical representation of the beats per minute as a function of time. The processing programming code does not do any computations for the pulse graph. All operations are performed on the Arduino board, thereby necessitating the use of Arduino programming language in order to interface with the graphical user interface (GUI). The programme functions by extracting data from the serial port and presenting it in a visual format.

Figures 5.5 and 5.6 illustrate the pulse wave seen on two distinct software platforms, namely the Java processing graphical interface and the serial plotter inside the Arduino IDE. Facilitating the simultaneous functioning of both displays simplifies the troubleshooting process during the development of the pulse detection algorithm. However, the serial plotter is limited to the presentation of pulse waves or photo plethysmographs. The pulse sensor is often used in medical devices for the purpose of non-invasive heart rate monitoring. Photo plethysmographs may or may not be used for the measurement of blood oxygen levels, often known as SpO_2. The sensor used in our study functions by amplifying the raw data and then standardising the pulse wave around V/2, which represents the midway point in voltage. The pulse sensor is capable of detecting and responding to variations in light intensity that are relative in nature. An increase in light intensity results in an elevation of the signal, whereas a fall in light intensity leads to a reduction in the signal. The intensity of the green LED light that is reflected back to the sensor undergoes changes with each pulse.

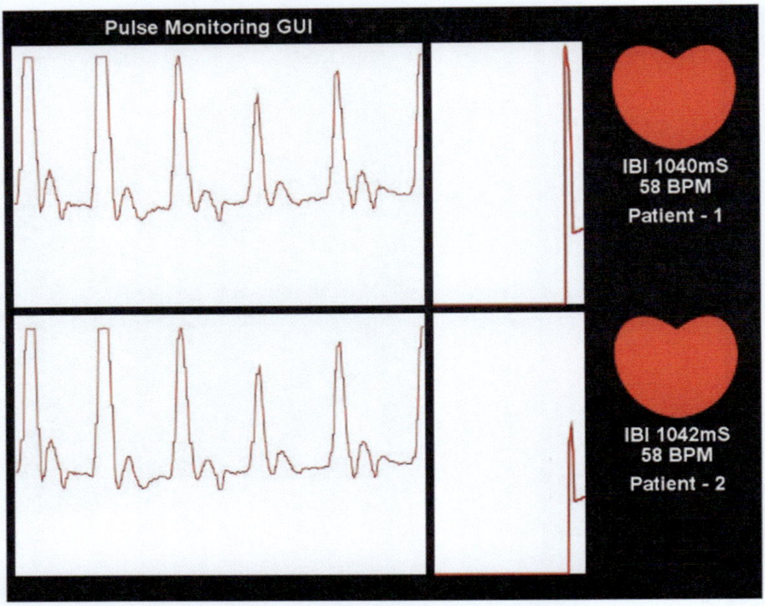

Fig. 5.5 UI interface

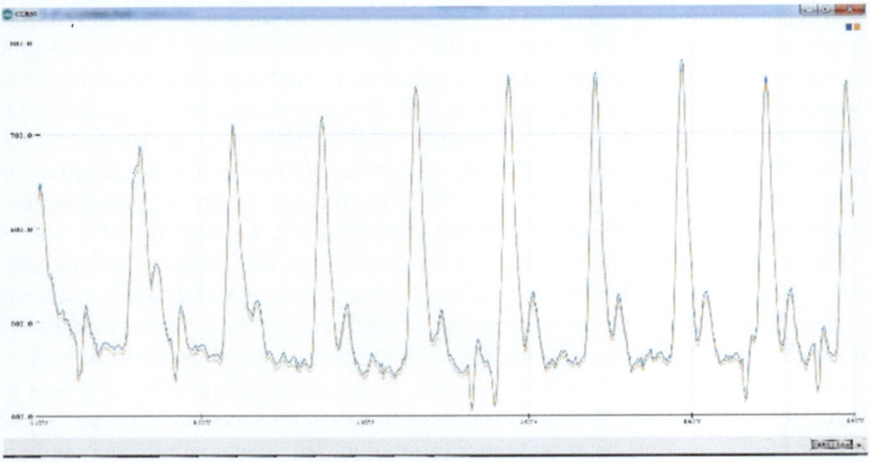

Fig. 5.6 Pulse wave in Arduino serial plotter

Table 5.1 Comparison between oscillometric and oximetric devices

No of people	Oscillometric device (BPM)	Oximetric (BPM)	Error state (%)
1	76	78	2.56
2	78	78	0
3	76	72	5.56
4	82	84	2.38
5	83	84	1.19
6	85	90	5.56
7	77	84	8.33
8	79	84	5.96
9	89	96	7.29
10	88	90	2.22

5.4 Result and Discussion

The performance of the oximetry device is tested with the output of oscillometric for 10 random people. The error rate is calculated using error rate formula [4]:

$$E = [100 \times |A - M|] \div A$$

where A = Actual Heart Rate (Oscillometric), M = Measured Heart Rate (Oximetric), E = Error Rate.

Table 5.1 shows the comparison data of pulse detected between the oscillometric and oximetric devices. Based on Table 5.1, the data obtained from the oximetric device has high accuracy with a mean of 4.31 and standard deviation of 2.87.

The results indicate a little margin of error, which is averagely below than 5% and hence deemed acceptable. The mistake in the experiment may be attributed to the movement of individuals throughout the process of obtaining readings. According to the findings of the study, it is recommended that the placement of the photosensor be chosen in areas with a high concentration of capillary blood in order to enhance the accuracy of heart rate calculations.

5.5 Conclusion

In conclusion, this study demonstrates the efficient and cost-effective real-time monitoring of heart rate using the proposed technology. This method eliminates the need for time-consuming and costly clinical pulse detection equipment. The graphical user interface is a convenient and user-friendly interface that is well-suited for medical applications. It combines both analogue and digital signal processing methods to maintain simplicity and effectively mitigate signal disturbances. The device might be seen as a rudimentary foundation upon which a diverse range of further applications can be incorporated. The sensor should possess enhanced power in terms of photo diode capability in order to provide improved detection capabilities, even in scenarios where blood capillary presence is minimal.

References

1. N.M.Z. Hashim, M.S. Sizali, Wireless patient monitoring system. IJSR **2**(8), 250–255 (2013)
2. R.G. Landaeta, O. Casas, R.P. Areny, Heart rate detection from plantar bioimpedance measurements, in *28 th IEEE EMBS Annual International Conference*, USA, (2006), pp. 5113–5116
3. N. Constant, T. Wang, K. Mankodiya, Pulseband: a hands-on tutorial on how to design a smart wristband to monitor heart-rate. Department of Electrical, Computer, and Biomedical Engineering University of Rhode Island, Kingston, RI, USA, (2015)
4. N.I. Ramli, M. Youseffi, P. Widdop, Design and fabrication of a low cost heart monitor using reflectance photoplethysmogram, WASET **08**, 417–418 (2011)
5. H. Shim, J.H. Lee, S.O. Hwang, H.R. Yoon, Y.R. Yoon, Development of heart rate monitoring for mobile telemedicine using smartphone, in *13th ICBME*, Singapore (2008), pp. 1116–1119

Chapter 6
Integration of Vehicle Safety and Surveillance System Using the Internet of Things Concept

Syahmi Izzat Bin Mohd Shafei, Mohd Fauzi Bin Alias, Mohamad Rosyidi Ahmad, Mohamed Yusof Radzak, Mohamad Shukri Mohd Zain, and Mohd Suyerdi Omar

Abstract Nowadays, motor vehicles are one of the objects that are very easy to steal. Preventing theft is not enough just by installing locks, but it also needs to equip the vehicle with monitoring and operator systems. Such vehicles are equipped with GPS (global positioning system), which can be monitored remotely. Internet of things (IoT) devices will be equipped with applications that will be connected to the Blynk server and then installed on the vehicle. If the vehicle is turned on, then the system will send data on the vehicle from the GPS satellite to the Blynk server. The use of the Blynk server was chosen because it has comprehensive content for various web platforms and mobile devices. As a result, real time data from the vehicle received by the server will be displayed directly through the web application of the remote-control system and monitor. The project also focuses on the use of GPS for cars. The primary goal of this initiative is to track the location of the car using GPS

S. I. Bin Mohd Shafei (✉) · M. F. Bin Alias · M. R. Ahmad · M. Y. Radzak · M. S. M. Zain · M. S. Omar
Electrical, Electronic and Automation Section, Universiti Kuala Lumpur, Malaysian Spanish Institute, Kulim, Kedah, Malaysia
e-mail: syahmi.shafei@unikl.edu.my

Mechanical Section, Universiti Kuala Lumpur, Malaysian Spanish Institute, Kulim, Kedah, Malaysia

M. F. Bin Alias
e-mail: fauzialias@unikl.edu.my

M. R. Ahmad
e-mail: mrosyidi@unikl.edu.my

M. Y. Radzak
e-mail: myusofr@unikl.edu.my

M. S. M. Zain
e-mail: msyukrimz@unikl.edu.my

M. S. Omar
e-mail: msuyerdi@unikl.edu.my

T. A. Razak et al. (eds.), *IT Application for Sustainable Living 2*,
SpringerBriefs in Applied Sciences and Technology,
https://doi.org/10.1007/978-3-031-93198-7_6

through the IoT system. Another objective is to be able to detect the heat and speed of the car. Users can use smartphones to track the location, heat, and speed of the car from a distance anytime and anywhere.

Keywords Speed sensor · GPS tracking system · Dht 11 sensor · IoT

6.1 Introduction

The automotive industry consists of a broad range of organizations and companies with a critical objective of developing, designing, manufacturing, marketing, and selling motor vehicles. By revenue, the automotive industry is an important portion of the world's economic sectors and vital to the economies of all major countries. The industry is still expanding, with a 30% increase over the last decade (1995–2005) [1]. Internet of things (IoT) devices installed in vehicles use GPS, RFID, and OBDII sensors to collect data. This data comprises the idling time, vehicle speed, location, load, driver behavior, fuel consumption, and temperature monitoring. A linked car is another type of IoT gadget that can communicate with other smart devices on the same network. With recent advancements in technologies such as wireless distributed sensors, embedded systems, cloud computing, light weight material, smart cognitive radio networks, higher efficiency, high performance battery, ultra-low emission internal combustion engines, intelligent converter, and fuel cell technology the production of smarter, safer, energy-efficient, and zero-emission vehicles are now possible.

6.1.1 Speed Sensor

Rotational speed measurements are an essential component in many industries, including automation and automotive. Position sensors and rotational speed, particularly in electrical drive systems are critical for overall machine functionality [1]. Sensors having digital outputs are favored for electrical drive and control systems as the industry digitalizes. On the market, there are multiple types of digital rotational speed and position sensors (see Fig. 6.1) such as inductive sensors, magnetic gear tooth sensors, encoders, and optical grating sensors. They can be used for a variety of applications depending on their structures and physical principles [2, 3].

Based on the goal of developing a new digital signal processing technique and the interoperability with current sensors such as the Hall-Effect gear tooth sensor [3], the following strategy has been proposed. The analogue periodic output signal of a rotating speed and position sensor was processed by utilizing analogue and digital signal processing techniques, as shown in Fig. 6.2.

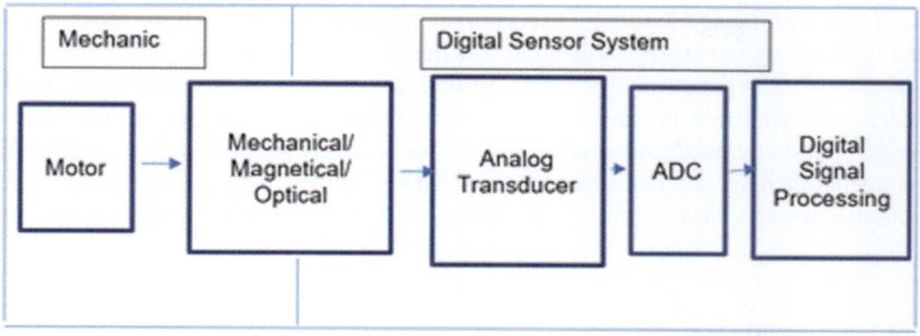

Fig. 6.1 Typical digital rotational speed sensor construction

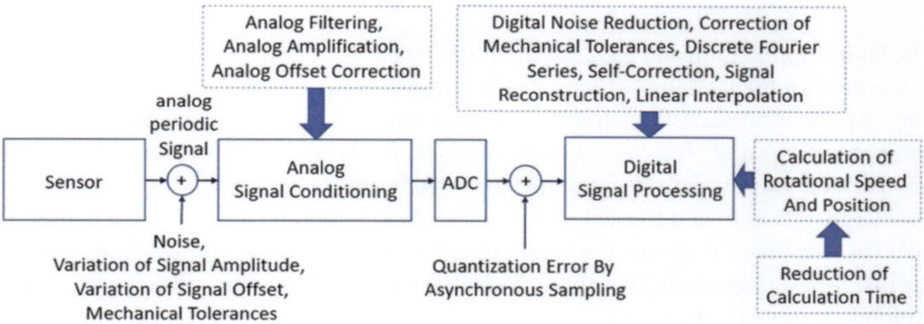

Fig. 6.2 Typical analog rotational speed sensor construction

6.1.2 GPS Module

GPS technology, also known as a satellite navigation system, transmits time, speed, and position information to the GPS receiver. It determines the latitude and longitude of a vehicle at any place in the world. Additionally, the device gets data about a car's physical location on Google Maps. As a result, it may be used everywhere on the earth's surface, regardless of the time or weather conditions [4]. Thus, car owners can track their vehicles in real time.

6.1.3 Internet of Things

The internet of things (IoT) refers to a network comprising interconnected computing devices, digital machines, mechanical objects, individuals, or animals. These entities possess unique identifiers (UIDs) and possess the capability to transmit data through a network without requiring direct human-to-human or human-to-computer interaction [5]. The internet of things encompasses a wide range of entities, including individuals equipped with heart monitors, farm animals implanted with biochip transponders, and vehicles equipped with built-in sensors that notify the driver of low tyre pressure. These entities, whether natural or man-made, are capable of being assigned an IP address and transmitting data over a network. Organizations across several sectors are progressively using the internet of things to enhance operational efficiency and get deeper insights into consumer behaviors. This enables them to provide enhanced customer service, make informed decisions, and generate more corporate value [6].

6.2 Methodology

6.2.1 Project Design

The Blynk application and the Arduino IDE server have been employed as the servers in the development of this system. The Blynk program also has a database architecture. Application servers or traditional programming languages are used to create computer programs. The period of project development will be broken up into iterations, with each iteration including the production of new modules with distinct characteristics assigned to each module in turn. The module that a person with their own car uses most frequently will be sought as a high priority during development, for instance, the module for vehicle location. The combination of modules will run on final increments once each module has been fully built in order to form a whole system. The project design is shown in Fig. 6.3.

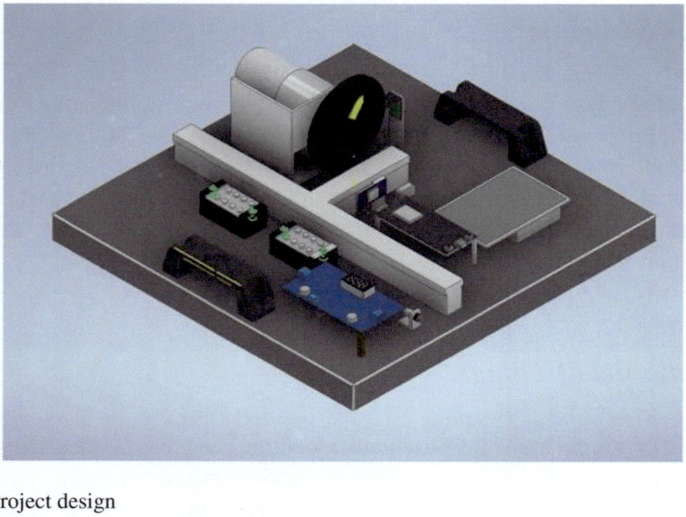

Fig. 6.3 Project design

6.2.2 Process Flowchart

Figure 6.4 shows a flow chart for the process. The first step is to obtain the exact location of the vehicle. After obtaining the information, the location's information of the vehicle will be retrieved through the GPS module. The system will send the location of the vehicle through the Blynk application available on the user's mobile phone. In addition, the system will also retrieve the temperature and the current speed values of the vehicle. In order to transfer the vehicle's information to a smartphone successfully, the internet or WIFI intermediaries are used. After connecting the WIFI terminal with the system, the user has to press the 'Play' button on the Blynk app and it will continue to show the ongoing results such as location, temperature, and vehicle's speed. When the 'Stop' button is pressed, the application will stop immediately. Figure 6.5 shows the Blynk and microprocessor systems integration.

6.2.3 Blynk and Microprocessor Systems Integration

See Fig. 6.5.

6.3 Result and Discussion

6.3.1 Blynk Platform (IoT Concept) and Microprocessor Systems

Figure 6.6 shows the results obtained from the application on the user's smartphone using the Blynk platform which also works as an IOT system. The figure shows the exact location obtained by displaying latitude and longitude on the platform. In addition, the Blynk application also displays the result of the temperature and humidity inside the vehicle. The Blynk application also displays the current speed of the vehicle.

Figure 6.7 shows the microprocessor system including LCD display and DC motor (for real time testing) that been placed inside the vehicle. The complete system implements a single unit of NodeMcu/esp8266 as its main microprocessor. The GPS sensor, the temperature and humidity sensor have been connected to this microprocessor. The system has implemented two others microprocessor as secondary units to the system named Arduino Uno and Mega. Arduino Uno was used for speed sensors connection and integration. Arduino Mega was used by the TFT LCD to display real time data.

Figure 6.8 shows the prototype of project. In the testing phase, the unit systems and user interface tests were conducted. The unit system testing is the functional testing of a single module, while user interface testing is testing for the presence of defects in the system's user interface. All systems were tested to find errors and

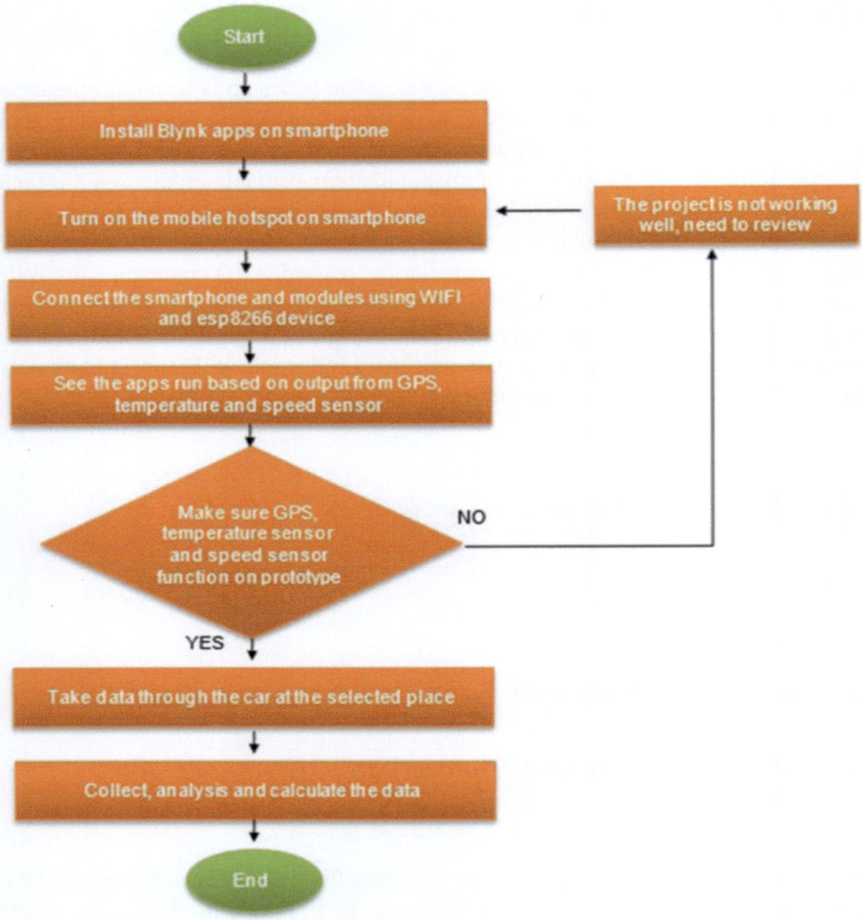

Fig. 6.4 Process flowchart

verify that the functionality works as expected by the user. Figure 6.8 also shows the results were successfully displayed on the TFT LCD which are latitude and longitude, temperature, and speed.

6.3.2 Data Collection and Graph Analysis

The testing result displays the data taken from the location of study near Kulim, Kedah to Sungai Dua, Pulau Pinang which covers a distance more than 24 km. The

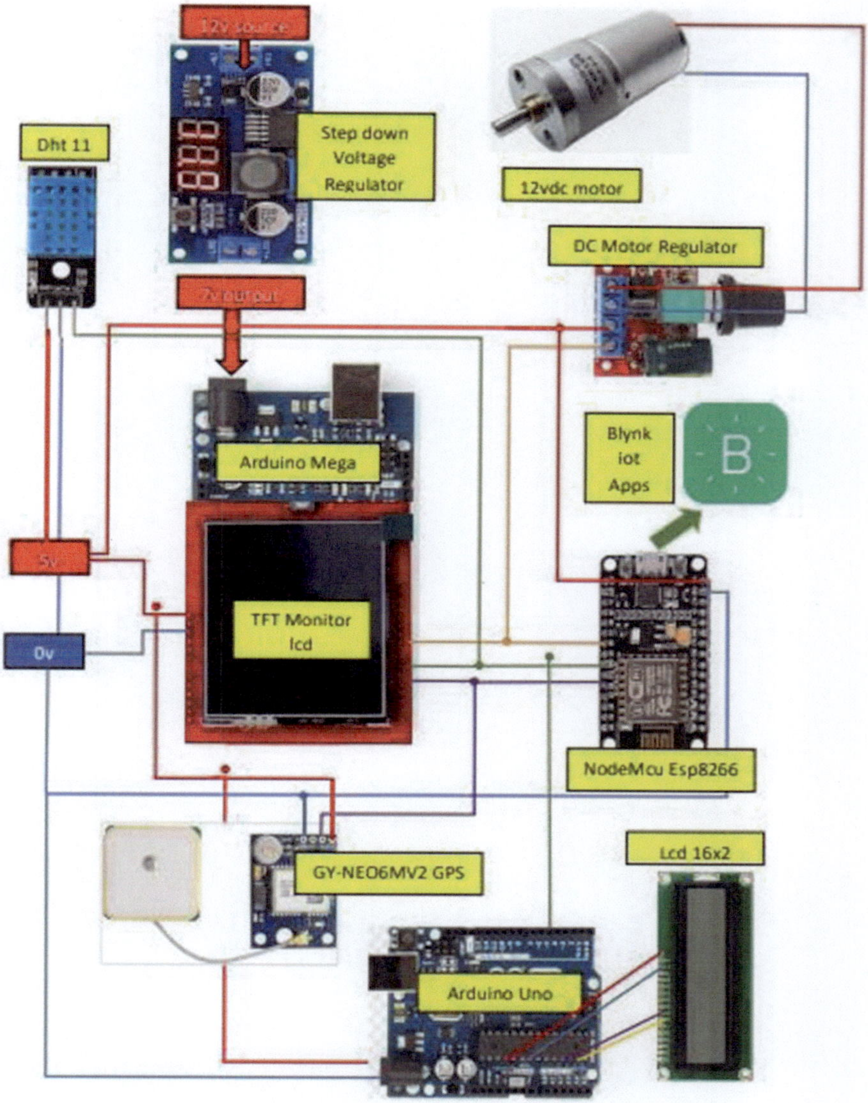

Fig. 6.5 Connection between Blynk and microprocessor

latitude and longitude data also were obtained at a location point in Kulim and Sungai Dua. The result is shown in Fig. 6.9.

Table 6.1 shows the speed, temperature and humidity data obtained from several tests that had been done using various car's speeds of 20, 40, 60, 80 and 100 km/h.

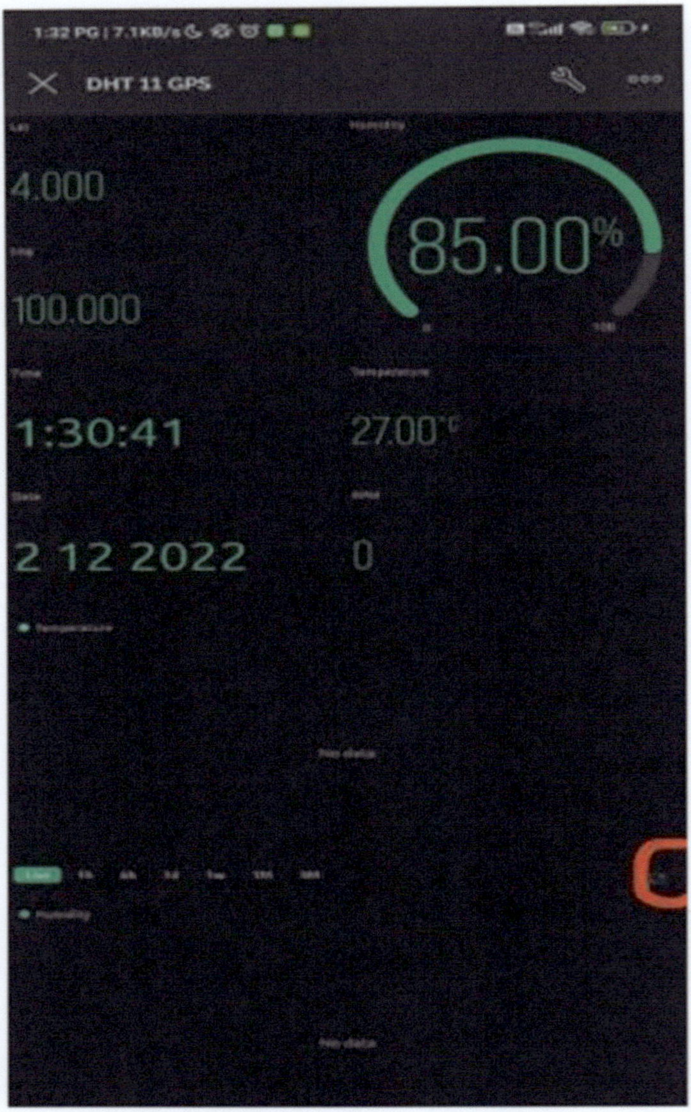

Fig. 6.6 Blynk result

The results show that the time taken is different and the temperature and humidity in the car are also different at different speeds.

Figure 6.10 shows the relation between the speed and the time needed for a single vehicle to move from two location points at Kulim and Sungai Dua. All the speed data were obtained through the Blynk display throughout the experiment. It shows

Fig. 6.7 Microprocessor system

similarity between the data obtained by the system and the real data from the vehicle's speedometer. As conclusion, the system has successfully displayed the accurate result compared to the real data. For temperature and humidity also apply to the weather level in the chosen place.

6.4 Conclusions

As a conclusion, the project has developed a successful system to locate the vehicle's location using a GPS component. The location has been displayed using longitude and latitude data. On the advantages side, the project has also successfully embedded the system with the speed, temperature, and humidity data. According to the experimental findings presented, all the data obtained from the system has been compared to real data and it shows the similarity between both data. Lastly, the system is also affordable and cost effective. Moreover, the system has a user friendly design which helps the user to access and use.

Fig. 6.8 Prototype testing result

KULIM		
Latitude:	**5.38945**	GPS: **5° 23' 22.0200" & 100° 33'**
Longlitude:	**100.563431**	**48.3516" E.**
SUNGAI DUA		
Latitude:	**5.370034**	GPS: **N 5° 31' 29.37", E 100° 27'**
Longlitude:	**100.413948**	**3.2868"**

Fig. 6.9 Latitude and longitude obtained

Table 6.1 Speed, temperature, and humidity result

Speed (km/h)	Theory time (min)	Experiment time (min)	Temperature (°C)	Humidity
20	72	82	84	24
40	36	45	84	24
60	24	35	86	25
80	18	26	88	25
100	14.4	20	90	26

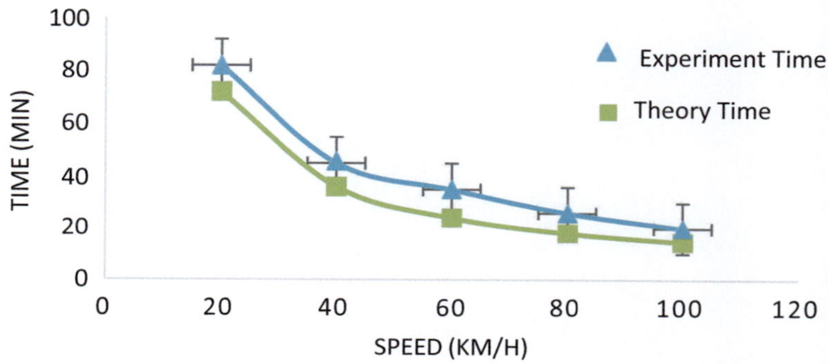

Fig. 6.10 Speed versus time

References

1. O.T. Arulogun, O. Omidiora, M. Olaniyi, A.A. Ipadeola, Development of security system using facial recognition. PJST, 377–386 (2008)
2. G. Divya, et al., Design and implement of wireless sensor street light control and monitoring strategy along with GUI. IJITR, 78–81 (2016)
3. S. Iranpak, A. Shahbahrami, H. Shakeri, Remote patient monitoring and classifying using the internet of things platform combined with cloud computing. J Big Data, 2021 (2021)
4. T.P. Huynh, Y.K. Tan, K.J. Tseng, Energy-aware wireless sensor network with ambient intelligence for smart LED lighting system control. IECON (2011)
5. A.A. Nippun Kumaar, G. Kiran, S. TSB, Intelligent lighting system using wireless sensor networks. IJASUC **1**(4), 17–27 (2010)
6. S.B. Song, W.Y. Cheon, Y.S. Yu, Design of a LED lighting bar replacement neon sign. JKIEE, 1671–1672 (2006)
7. X.Y. Wang, A.L. Fitch, H.H.C. Iu, V. Sreeram, W.G. Qi, Implementation of an analogue model of a memristor based on a light-dependent resistor 2012 CPS/IOP (2012)

Chapter 7
Low Harmonics Plug-In Home Charging Electric Vehicle Battery Charger Utilizing Multi-level Rectifier, Zero Crossing, and Buck Chopper

Halina Hassan, Norhalimatul Sadiah Kamaruddin, Amir Shauqee Abdul Rahman, Mohd Rezal Mohamed, Saharul Arof, Emilia Noorsal, Hamzah Arof, and Philip Mawby

Abstract A conventional rectifier generates a significant quantity of output harmonics current. This paper describes zero crossing control and the IGBTS firing sequence control technique, which are integral elements of the low harmonics plug-in home charging electric vehicle battery charger employing a multi-level rectifier, zero crossing, and buck chopper. Using MATLAB and Simulink, the zero-crossing control and IGBTS discharge sequence control techniques are studied and tested. The

H. Hassan (✉) · N. S. Kamaruddin · A. S. A. Rahman · M. R. Mohamed · S. Arof
Electrical, Electronic and Automation Section, Universiti Kuala Lumpur, Malaysian Spanish Institute, Kulim, Kedah, Malaysia
e-mail: halina@unikl.edu.my

N. S. Kamaruddin
e-mail: norhalimatul@unikl.edu.my

A. S. A. Rahman
e-mail: amirshauqee@unikl.edu.my

M. R. Mohamed
e-mail: mrezal@unikl.edu.my

S. Arof
e-mail: saharul@unikl.edu.my

E. Noorsal
Center for Electrical Engineering Studies, Universiti Teknologi MARA, Pulau Pinang, Malaysia
e-mail: emilia.noorsal@uitm.edu.my

H. Arof
Engineering Department, Universiti Malaya, Jalan Universiti, Kuala Lumpur, Malaysia
e-mail: ahamzah@um.edu.my

P. Mawby
School of Engineering, University of Warwick, Coventry, UK
e-mail: P.A.Mayby@warwick.ac.uk

T. A. A. Razak et al. (eds.), *IT Application for Sustainable Living 2*,
SpringerBriefs in Applied Sciences and Technology,
https://doi.org/10.1007/978-3-031-93198-7_7

MATLAB/Simulink results demonstrate that the proposed technique, when applied to the proposed battery charger, outperforms the conventional rectifier battery charger, in which the control signal draws pure sinusoidal line current. The proposed zero-crossing and IGBT discharge sequence is ideally suited for a low-harmonics plug-in home charging electric vehicle battery charger employing a multi-level rectifier, zero-crossing, and buck-chopper for the electric car's battery charger.

Keywords EV · Battery charger · THD · IGBT

7.1 Introduction

Government, industry, and academia are collaborating to create an electric vehicle (EV)-based transportation system that is connected to the grid and will significantly reduce the consumption of fossil fuel in response to growing concerns about the negative effects on the environment caused by our daily transportation, which is responsible for increasing greenhouse gas emissions around the globe.

Using efficient electric vehicles is one of the solutions to reduce global hydrocarbon emissions [1]. The batteries of electrically powered vehicles must be charged to restore their energy after use. It can be replenished by converting the chemical energy of the internal combustion or the kinetic energy of the moving vehicles into electricity.

A new series-motor four quadrants DC chopper (FQDC) with multiple operations was designed for DC drive motor that can be used for motor traction for DC EV [1]. The FQDC can overcome two common problems with series motor which are speed drop when loaded and overrun when running at low load. The FQDC was designed for the application of a DC drive electric car which is economical as an alternative to AC drive electric car. The research works of FQDC can be segregated as FQDC optimization and chopper operation optimization using artificial intelligence (AI), and EV battery charger. The proposed electric car home charging battery charger to be associated to the FQDC consists of zero crossing (ZC), multi-level rectifier (MLR) and buck chopper (BC) [2–6]. Those combinations require control and sequencing algorithms in controlling the battery charger and are described in detail in [3, 7, 8]. The charging action requires battery state of charge (SOC) known for charging voltage and current to guarantee the safety and life of the batteries as discussed in [4, 5]. A battery charger consists of a transformer, a rectifier, and a buck or boost chopper. It should be portable, high in power density and cheap. The battery charger and its control algorithm are implemented using microcontrollers, digital signal processors, and integrated circuits [7]. Along with charging time and battery lifetime, decrease in the power factor due to an increase in the firing angle and relatively high harmonic are critical aspects to consider when designing a battery charger. Thus, multi-level converters are introduced to attain high power quality, low switching losses, high voltage capability, and better efficiency [2, 9–12]. It improves the power factor and reduces the total harmonic distortion (THD) [6]. The proposed battery charger requires zero crossing [6, 7, 9, 10] and IGBTs firing circuits to generate the IGBTs firing for the multi-level rectifier. This paper aims to design IGBTs firing signals for the proposed battery charger to be applied for home charging EV battery charger.

7.2 Methodology

A MLR has several IGBTs in its circuit. The MLR requires IGBTs to generate correct and precise firing signals for the incoming AC voltage supply to sustain its performance. Details on how the firing sequence is generated and linked to zero cross are discussed next. The MATLAB/Simulink software is used for the circuit design and to simulate the circuit.

7.2.1 IGBT Firing Sequence Pulse Detection

The firing sequence and the process for obtaining the IGBT firing can be broken into separate subsequences. The first subsequence is shown in Fig. 7.1. As can be seen, the triangular signal is used as part of the signal conversion process. Vs is the incoming AC voltage supply from the grid. The offset of the incoming voltage grid supply is converted to absolute sine and the triangular signals as shown in Fig. 7.2.

An absolute signal is produced by the negative part of the sine wave signal and is converted to the positive side. Offset means the generated absolute signal is lowered

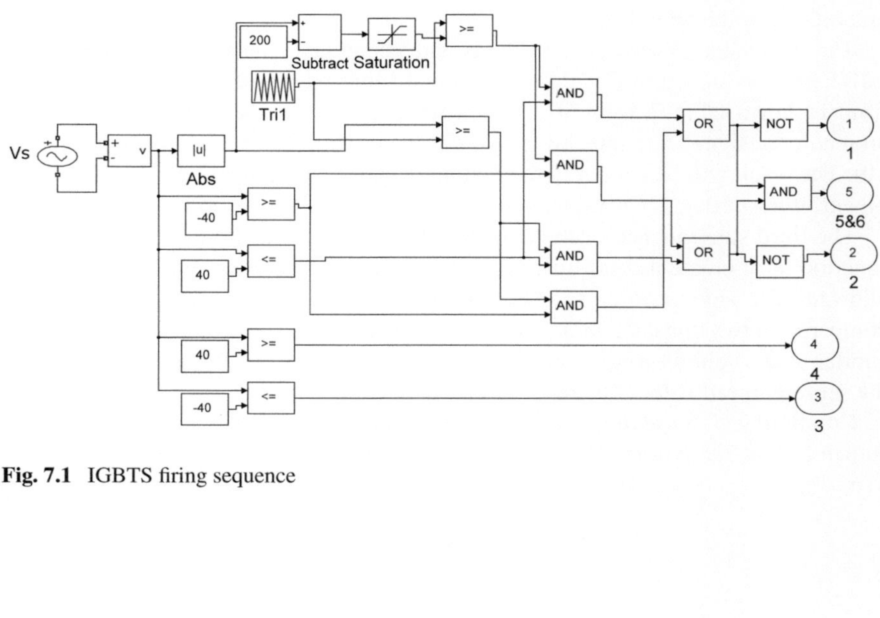

Fig. 7.1 IGBTS firing sequence

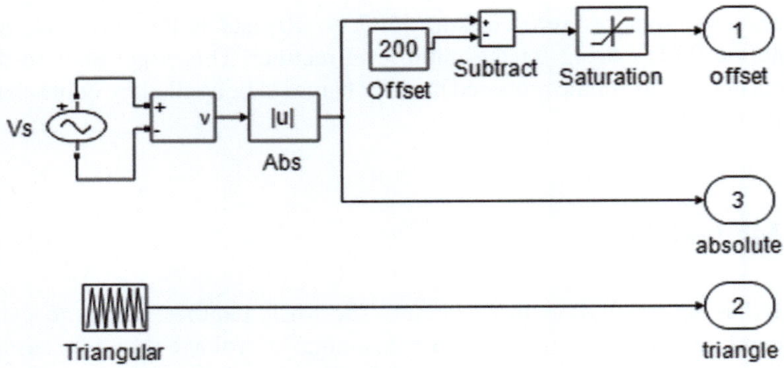

Fig. 7.2 First subsequences

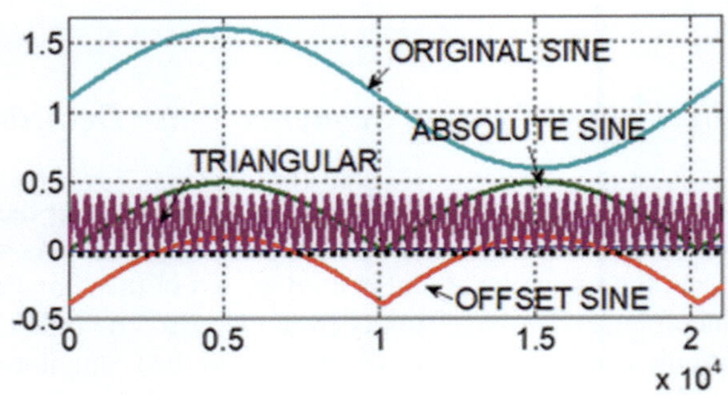

Fig. 7.3 Signals result of subsequence

by doing some math subtraction with a constant value to offset it. The absolute signal and offset are shown in Fig. 7.3.

The final value of the offset, absolute and triangular signals used to generate the IGBT firing is shown in Fig. 7.4. The second subsequence of the complete IGBTS firing is shown in Fig. 7.5 and 7.6. The sine wave signal from the grid which represents an incoming signal is compared to two constant values (in this example, 10 and − 10). The result will be similar to the Schmitt trigger output function '1' or '2' if it is bigger or lower than the set referenced value and the result is shown in Fig. 7.6.

The third subsequence of the complete IGBTS firing is shown in Fig. 7.7. The function is to create a dead time zone. In this dead time zone, no IGBT firing is allowed. The sine wave signal from the grid representing an incoming signal is compared to two constant values (in this example, 40 and − 40). The result will be similar to the Schmitt trigger output function, '2' or '1' if it is bigger or lower than the set referenced value. The result is shown in Fig. 7.8.

Combining all these three subsequences will result in the IGBT firing signals, as shown in Fig. 7.9. Number 1, 2, 3, 4, 5, and 6 represent the firing signal for IGBT1 to 6.

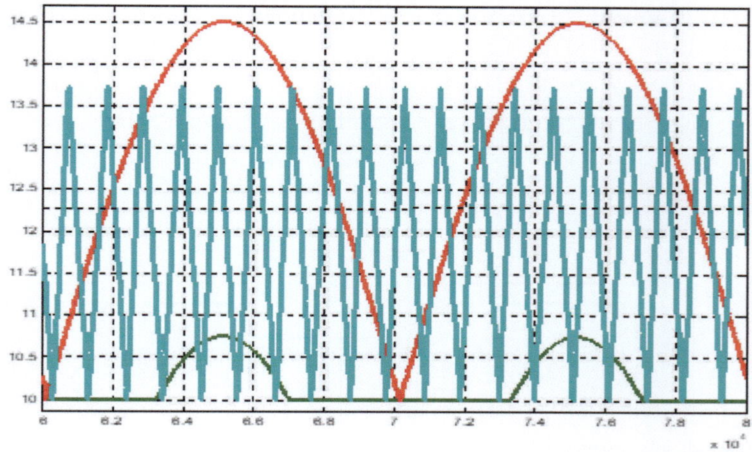

Fig. 7.4 Final result subsequence 1

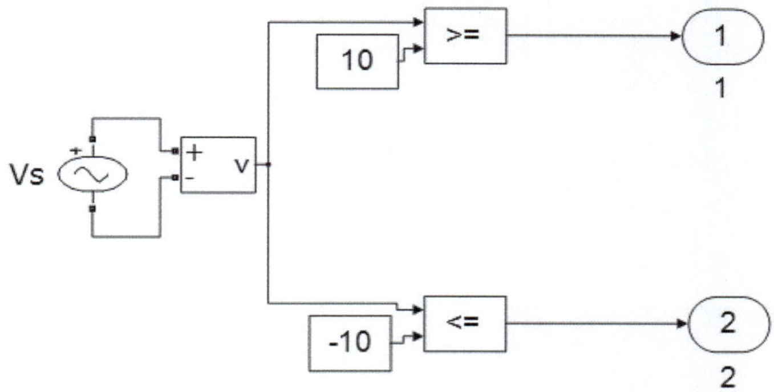

Fig. 7.5 Subsequence 2

The complete combinations of IGBT firing and the zero cross is as shown in Fig. 7.10.

7.3 MLR Simulation Results with MATLAB Simulink

A simulation model of MLR [6, 7] is used to compare the performance of a proper zero-cross assist IGBT firing with the wrong/false signals. The simulation model investigates the performance of the battery charger, shown in Figs. 7.11, 7.12, 7.13, 7.14 and 7.15. If the sequence of zero cross and IGBTS firing signals are correct, the output of the MLR will be as in Figs. 7.14 and 7.15. Else the result will be as

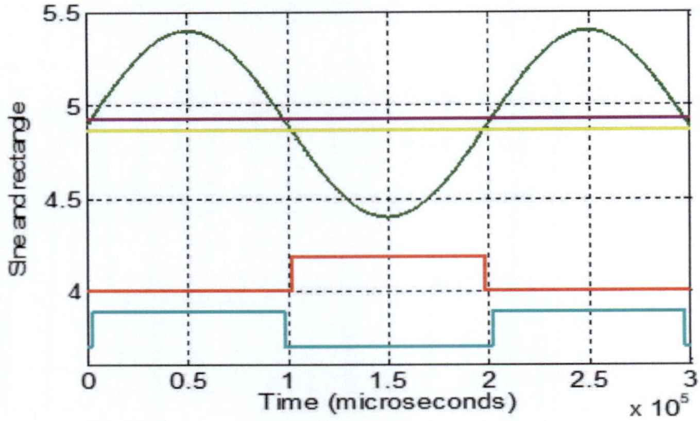

Fig. 7.6 Result subsequence 2

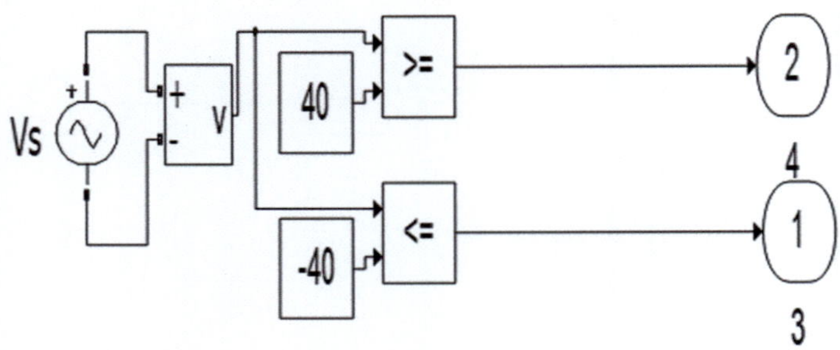

Fig. 7.7 Subsequence 3

Fig. 7.8 Result subsequence 3 dead time zone

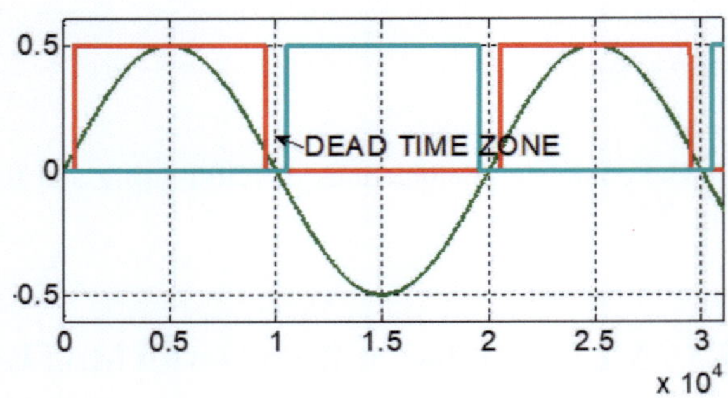

in Figs. 7.11, 7.12 and 7.13. The incorrect voltage Vab of the MLR is, as shown in Fig. 7.11.

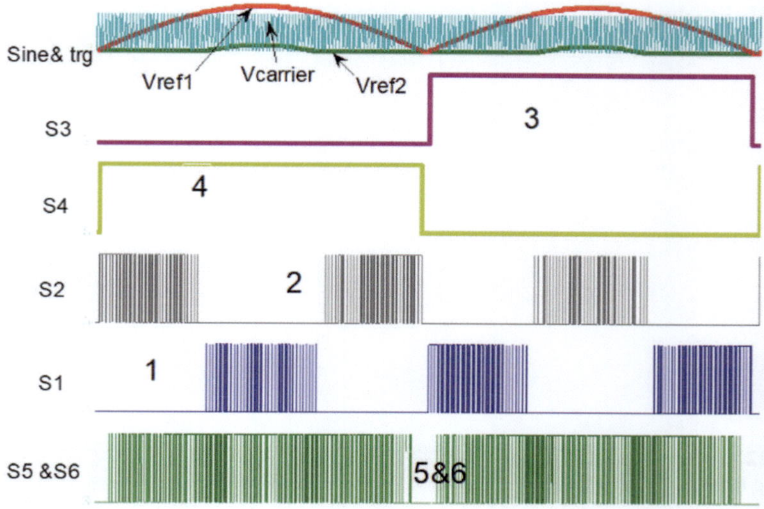

Fig. 7.9 IGBTS firing signal

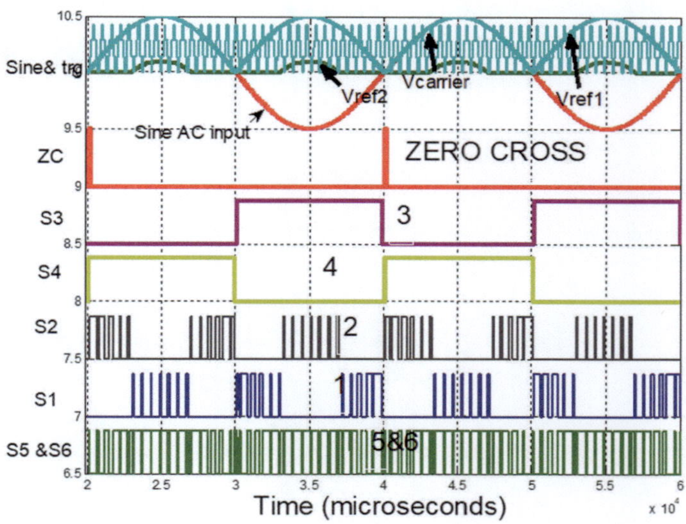

Fig. 7.10 ZC and input and output and IGBT firing signals

The incorrect voltage Vab of the MLR is shown in Fig. 7.12 as well. The incorrect input current of this multi-level inverter is, as shown in Fig. 7.13. The correct voltage Vab of this multi-level inverter is, as shown in Fig. 7.14.

The correct input current of this multi-level inverter is, as in [6]. The correct output voltage of MLR of this multi-level inverter is as shown in Fig. 7.15. If the zero-cross

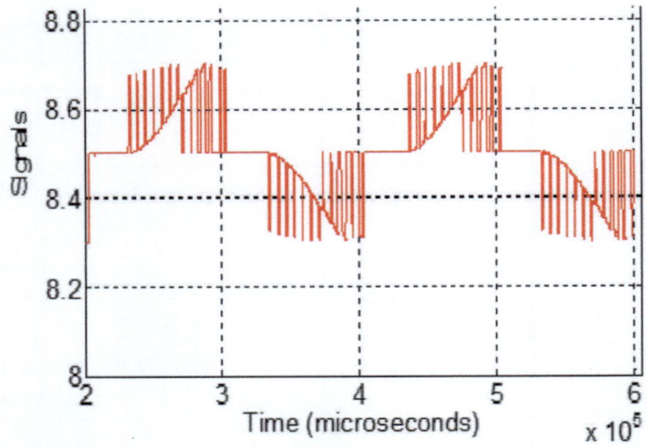

Fig. 7.11 False/wrong Vab voltage of MLR

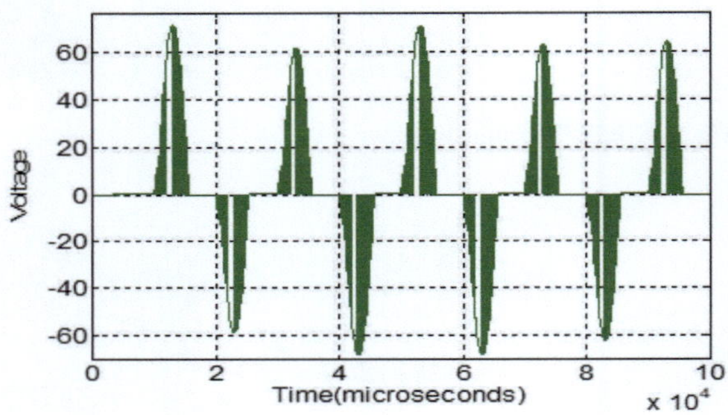

Fig. 7.12 Voltage Vab MLR type flip firing

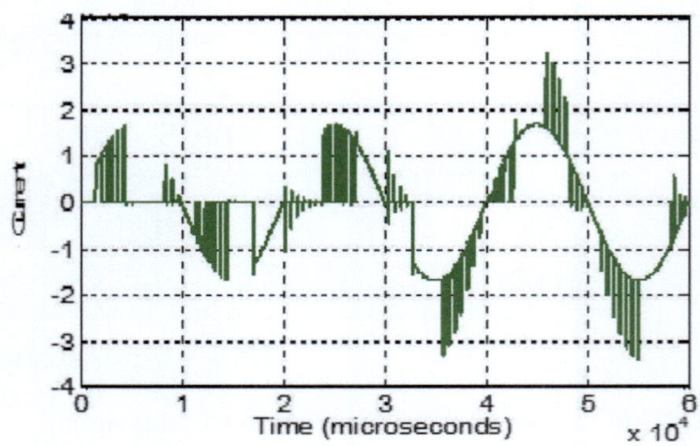

Fig. 7.13 Input current wrong firing

and the IGBTs firing is correctly fired and synchronized, the MLR will result in low THD [6].

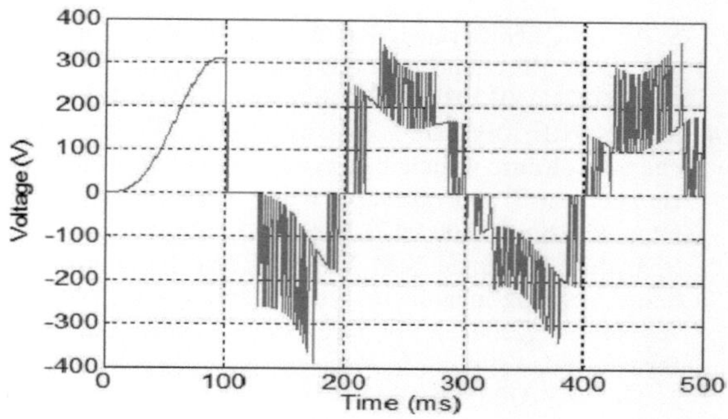

Fig. 7.14 Output voltage Vab multi-level type

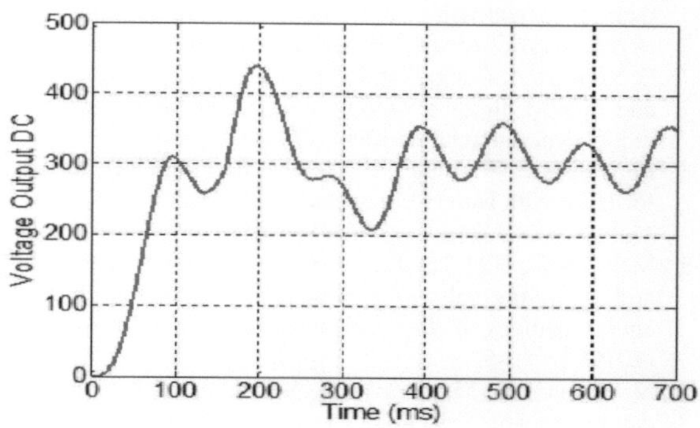

Fig. 7.15 Voltage output of MLR

7.4 Conclusion

This paper discusses about the IGBT firing and sequences control which is crucial to assure the successful operation of the multi-level rectifier. The proposed battery charger circuit using multi-level rectifier with proposed IGBT firing technique shows a better THD, fast response, less ripple, and is suitable for battery charger for home charging EV. The proposed technique is successfully tested and is ideal to be linked to the zero-cross and IGBT firing technique for the proposed battery charger.

References

1. S. Arof, J.A. Jalil, N.M. Yaakop, P.A. Mawby, H. Arof, Series motor four quadrants drive dc chopper part 1: overall, in *International Conference on Power Electronics* (2014)
2. S. Arof, N.H.N. Diyanah, P.A. Mawby, H. Arof Emilia Noorsal, Series motor four quadrants drive DC chopper part 3: field weakening mode. IOP Conf. Ser. Mater. Sci. Eng. **1127**(1), 012028 (2021). https://doi.org/10.1088/1757-899X/1127/1/012028

3. S. Arof, N.H.N. Diyanah, P.A. Mawby, H. Arof Emilia Noorsal, New series motor four quadrants drive DC chopper for economical EV, part5: parallel mode', IOP Conf. Ser. Mater. Sci. Eng. **1127**(1), 012025 (2021). https://doi.org/10.1088/1757-899X/1127/1/012025
4. S. Arof, N.H. Diyanah, P. Mawby, H. Arof, N. Mohd Yaakop, Low harmonics plug-in home charging electric vehicle battery charger utilizing multi-level rectifier, zero crossing and buck chopper: part 1: general overview, in Progress in Engineering Technology (2019), pp. 103–108. https://doi.org/10.1007/978-3-030-28505-0_9
5. S. Arof, M.R. Ahmad, N.M. Yaakop, P. Mawby, H. Arof, E. Noorsal, A.A. Bakar, Y.M. Ali, Low harmonics plug-in home charging electric vehicle battery charger utilizing multilevel rectifier, zero crossing, and buck chopper: state of charge estimator using current integration algorithm for embedded system, in *Progress in Engineering Technology III* (2021). https://doi.org/10.1007/978-3-030-67750-3_19
6. M.S. Saharul Arof, N.M. Sazali, J.N. Noor, P.M. Amirah, E. Noorsal, A.A. Bakar, Y.M. Ali, Low harmonics plug-in home charging electric vehicle battery charger utilizing multi-level rectifier, zero crossing, and buck chopper: BMS battery charging control algorithm, in *Progress in Engineering Technology III* (2021). https://doi.org/10.1007/978-3-030-67750-3_18.
7. E. Noorsal, A. Rongi, I.R. Ibrahim, R. Darus, D. Kho, S. Setumin, Design of FPGA-based SHE and SPWM digital switching controllers for 21-level cascaded H-bridge multilevel inverter model. Micromachines **13**(2), 179 (2022)
8. L. Wenshan, W. Xuhui, P. Meng, Charging current control strategy during mode switching for HEV with battery connected to DC bus directly, in *International Conference on Electrical Machines and Systems (ICEMS)* (2018), pp. 974–977
9. M.R. Islam, H.H. Lu, M.J. Hossain, L. Li, Compensating neutral current, voltage unbalance and improving voltage of an unbalanced distribution grid connected with EV and renewable energy sources, in *International Conference on Electrical Machines and Systems (ICEMS)* (2019), pp. 1–5
10. N. Sivakumar, M. Saravanan, S.C. Raja, S.A.E. Xavier, Analysis of the performance of energy transfer system (ETS) for EV battery charging method using non-isolated SEPIC converter-based CCCV control, in *International Conference on Electrical Machines and Systems (ICEMS)* (2023), pp. 337–381
11. M. Reza Pazhouhandeh, M.M. Chang, T.A. Valiante, R. Genov, Track-and-zoom neural analog-to-digital converter with blind stimulation artifact rejection. IEEE J. Solid-State Circ. **55**(7), 1984–1997 (2020)
12. X. Tang, et al., An energy-efficient Time-domain incremental zoom capacitance-to-digital converter. IEEE J. Solid-State Circ. **55**(11), 3064–3075 (2020)

Chapter 8
Expert System and Direct Torque Control to Extend the Electric Braking at the End of Resistive Braking Mode

Saharul Arof, Norramlee Mohamed Noor, Mohamad Rosyidi Ahmad, Philip Mawby, Hamzah Arof, and Emilia Noorsal

Abstract Regenerative braking and resistive braking are the two most frequent types of braking used for electric vehicles (EVs). These types of braking are used to slow down the EV and recharge the batteries. When the motor speed is greater than its base speed, the regenerative braking system is activated first, followed by the resistive braking system. To bring an electric vehicle to a halt, regenerative and resistive brakes can only be effective up to a specific motor speed, and they have to be replaced with a mechanical brake that uses a pad. Because of the excessive braking torque that is created, it is usual for traditional choppers to suffer a jolt at the conclusion of resistive braking. This is because resistive braking generates a lot of force. Applying early mechanical brake intervention before the resistive braking enters this stage is the recommended strategy for overcoming this high torque using the traditional way. This results in the pad brakes wearing down faster than normal

S. Arof (✉) · N. M. Noor · M. R. Ahmad
Universiti Kuala Lumpur, Malaysian Spanish Institute Kulim Hi-Tech Park, Kulim, Kedah, Malaysia
e-mail: saharul@unikl.edu.my

N. M. Noor
e-mail: noramlee@unikl.edu.my

M. R. Ahmad
e-mail: mrosyidi@unikl.edu.my

P. Mawby
University of Warwick School of Engineering, Coventry, UK
e-mail: P.A.Mawby@warwick.ac.uk

H. Arof
Engineering Department, Universiti Malaya, Jalan Universiti, Kuala Lumpur, Malaysia
e-mail: ahamzah@um.edu.my

E. Noorsal
Faculty of Electrical Engineering, Universiti Teknologi MARA Cawangan Pulau Pinang, Permatang Pauh, Pulau Pinang, Malaysia
e-mail: emilia.noorsal@uitm.edu.my

T. A. A. Razak et al. (eds.), *IT Application for Sustainable Living 2*,
SpringerBriefs in Applied Sciences and Technology,
https://doi.org/10.1007/978-3-031-93198-7_8

and a loss of the electrical energy that was created, which might have been utilized to charge the auxiliary batteries. This work discusses the use of an expert system in conjunction with the direct torque control approach to prolong the electrical braking action at the conclusion of the resistive braking mode of a new four quadrants DC chopper. This approach is useful for rectifying a shorter resistive braking duration and excessive torque at the conclusion of the resistive braking mode in DC drive electric cars, which can be efficiently done with this method. The excess torque at the end of the electrical brake issue can be solved, according to the simulation results obtained using MATLAB/Simulink, which demonstrate that the expert system associated with direct torque control can do so.

Keywords Regenerative braking · Resistive braking · DTC · EC · FQDC · Expert system

8.1 Introduction

The use of electrical brakes has a sizeable influence on the deceleration of cars, most often electric vehicle kinds. The electrical braking can take the form of magnetic induction, the more frequent eddy current type, magnetic rhetological, regenerative, or resistive braking. It can also be of the magnetic rhetological type. The difference between regenerative braking and resistive braking is that the braking effect produced by regenerative braking is created by the motor itself, which does so by creating a motor torque in the opposite direction of the movement of the vehicle. When the motor is operating at a speed that is higher than its normal base speed, the regenerative braking action can be utilized, whereas the resistive braking action is utilized when the speed is lower than the nominal speed. When opposed to resistive braking, regenerative braking [1–3] enables battery charging and generates a greater voltage than conventional braking does. The braking effect is based on direct torque management, and the look up table (LUT) is often used to adjust the respected brake torque according to vehicle speed with a close loop feedback controller. The problem with these electrical braking systems is determining whether they will brake according to the desired braking effect without being either too forceful, which might cause the vehicle to jerk, or too sluggish, which will not be useful from the driver's point of view.

2009 was the year that saw the successful completion of the design of a DC brushed motor with high power output 55 kW, high efficiency (92%), and the ability to function at low operating voltage 13 V. As a reaction, a new four quadrants DC chopper (FQDC) [1] was developed and this motor is able to be linked to a supercapacitor, which enables rapid charging of electric vehicles (EV). This chopper allows for additional operations, such as driving, field weakening, parallel mode regenerative braking, and resistive braking [2], or it may enable motor reverse for EV applications. A look up table (LUT), which is constructed utilizing optimization tools such as genetics algorithm and gradient descent, is used [4, 5] for certain FQDC chopper operating modes. During the process of electrical braking, there are two different modes that may be employed. If the motor speed is above its nominal speed because of field weakening or parallel mode, it is most typical to begin the electrical braking process with regenerative braking and then proceed to resistive braking [6]. If the motor speed is below its normal speed, it is most common to begin the electrical braking process with resistive braking. The car will begin to slow down as a result of the regenerative braking system until it reaches the point when the regenerative braking system is no longer functional. Even while using regenerative braking, the motor continues to run at a speed that is higher than its normal speed. Batteries may be charged using the electricity that is generated by the generator. When the motor speeds drop below the nominal speed, resistive braking will take over and operate the vehicle. During this braking, the voltage of the produced power is lower than the voltage of the battery; as a result, the main battery is unable to be charged. However, any auxiliary battery that has a voltage that is much lower than the main battery may be charged. The resistive braking mode, which is shown in Fig. 8.1, has been extensively described in [1, 7, 8]. The motor or vehicle is brought to a stop using the electrical brake, which is the purpose of utilizing resistive braking. In the event that the regenerative braking system has a malfunction or malfunctioning components, the resistive braking system must be used so that the newly produced power may be dissipated. This backup alternative is essential.

In this study, the expansion of the resistive braking zone and the reduction of excessive braking torque at the conclusion of resistive braking mode operation of FQDC utilizing an expert system are the primary focuses of the investigation.

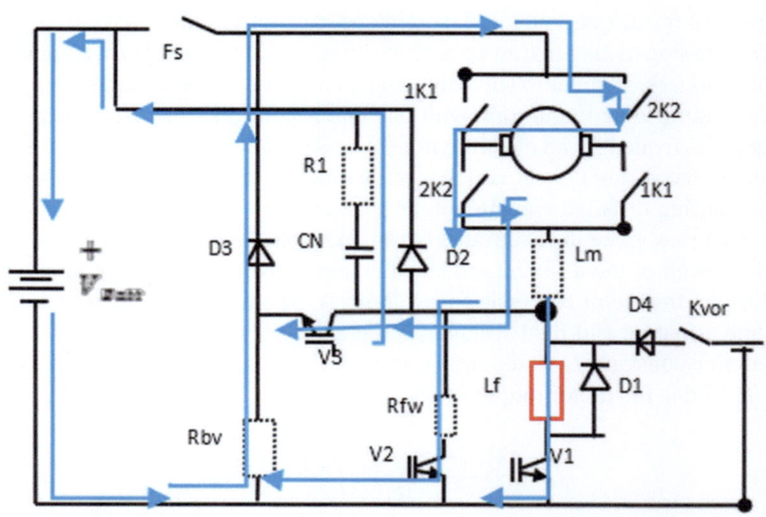

Fig. 8.1 IGBT V1 and IGBT V2 are switched ON

8.1.1 Direct Torque Control in Resistive Braking Mode

It is necessary to exercise control over the braking torque of the motor while using the resistive braking mode in order to achieve linear braking. It has been mentioned that the look up table and a direct torque control are available in [2]. However, in this work, it is employed for braking modes, while in [2], it is used for driving operations. However, one can still use the same LUT and direct torque control; the only difference is that it operates in reverse torque. The spindle speed motion and direction may be fought back against with the help of the negative torque. A comparison is made between the accelerator demand and either the vehicle speed or the motor speed to execute direct torque control. At each point in the speed spectrum, a specific amount of motor torque is associated with it. At each point where the signal from the brake pedal is being entered, the required torque is also being allocated to it. After contrasting the intended torque from the brake pedal and the speed, the number with the lowest difference is chosen as the final torque requirement. In order to have a linear braking action and eliminate any unexpected jerking, particularly while travelling at high speed, the needed brake torque should initially start at a low value. After then, the braking torque of the motor is raised linearly. Because of this effect, while the motor is operating at a high speed, the braking torque begins at a low value and gradually grows as the speed decreases, reaching a peak value when the motor is operating at a low speed. Figure 8.2, a simulation result of a 0.6 kW motor of DC series motor connected to FQDC in resistive braking mode, exhibits a condition if the braking torque is not controlled, it could lead to excessive braking torque, which occurs when the actual torque is greater than the reference torque. This excessive braking torque could lead to unintended results, such as the vehicle coming to a complete stop (acting like an emergency brake), or at the very least, it could cause the vehicle to jerk. In order to solve this issue, it is standard procedure to replace the resistive braking zone in question with one that is controlled by mechanical brakes [9], that are activated by an electrohydraulic brake. However, as a result of this, the hydraulic brake pad will wear and tear more quickly, and the electrical energy that is created during resistive braking will be squandered.

The equations given in (1) through (4) are examples of generic equations that can be applied to all of the different modes of chopper operation. If is the field current, *R*a and *R*f are the motor coil resistances, *B*emf is the back emf of the motor, *V*a and *V*f are the armature and field voltages, *K*b is the back emf constant, and *K*t is the torque motor constant. *I*a is the armature current; *w* is the angular speed, *J* is the inertia and *T*d is the motor torque.

$$I_a = \frac{V_{\text{batt}} - I_a\left(R_a + R_f\right) - B_{\text{emf}}}{L_a + L_f} \tag{8.1}$$

$$B_{\text{emf}} = K_v I_f \omega \tag{8.2}$$

$$T_d = K_t I_f I_a \tag{8.3}$$

Fig. 8.2 Torque control

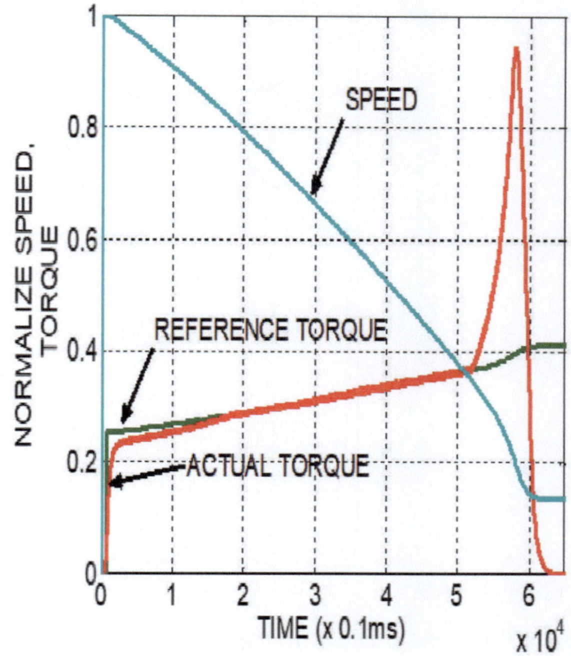

$$T_d = J\frac{dw}{dt} + B_w + T_L \tag{8.4}$$

Figure 8.2 shows that initially the motor speed fell in a linear fashion. Up to time 5, the torque is being managed such that it corresponds to the reference torque. After time 5, the torque is no longer controlled, which means that the resistive braking mode has ended. At this point, the mechanical brake should take over to bring the vehicle to a halt. In this study, we want to make the most of the electrical braking and the resistive braking for as long as possible before resorting to the electromechanical brakes. At the conclusion of the resistive braking mode, it is possible to see an increase in torque that is in excess of the reference point. This will result in an uncomfortable and unwelcome brake in the action.

Figure 8.3, a simulation of 0.6 kW of DC series motor connected to FQDC running resistive braking mode displays the consequence of an increase in the field current which causes a rapid decrease in the motor speed. As the speed drops due to braking effect in resistive braking mode, more current is needed to maintain the same generated voltage to charge batteries. According to Eq. (8.3), [1, 2], the field current has to be raised when the speed of the motor decreases in order to keep the same generated armature voltage. Because the same current that flew to the field winding originated from the armature current, an increase in the field current will also result in an increase in the armature current. Because both currents have already grown, the torque, which

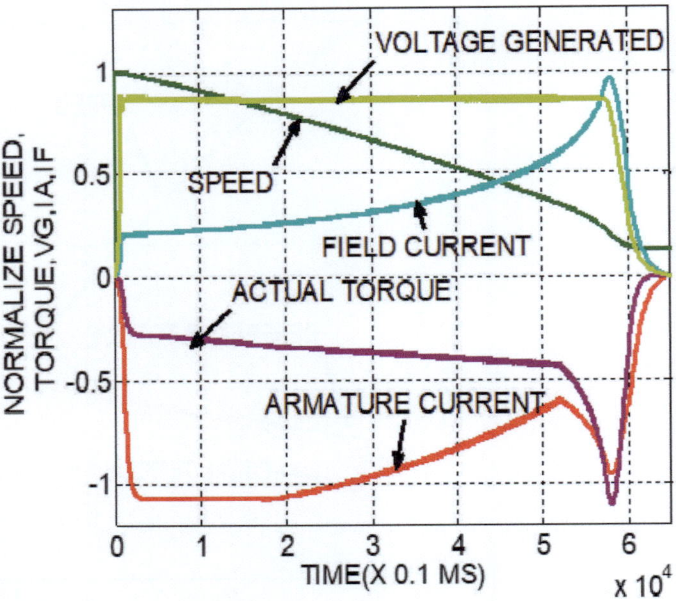

Fig. 8.3 Resistive braking current, speed, torque and voltage

is a factor of the field and armature currents, climbs above the necessary projected reference level which is unwanted.

8.2 Methodology

8.2.1 Expert System

An expert system [6, 10] which needs the expertise of an expert to handle the issue, is implemented as an if then rules algorithm to find a solution to this. At this point, the field current, which initially varied to create the needed amount of braking torque, must be maintained in order to keep the resistive braking torque at its original level. In its initial configuration, the system was armature voltage controlled; however, this has now been altered to field current control. The field current value that is ultimately selected is derived from the most recent reading taken before the battery charging voltage was switched from controlling the armature voltage to controlling the field current. An expert system method is also used to find out the condition, whether there was any excessive utilization of torque. Alternately, the speed at which the transition of torque from a condition which is controlled to become unmanageable may be set by the user in memory as part of the expert system algorithm. The field

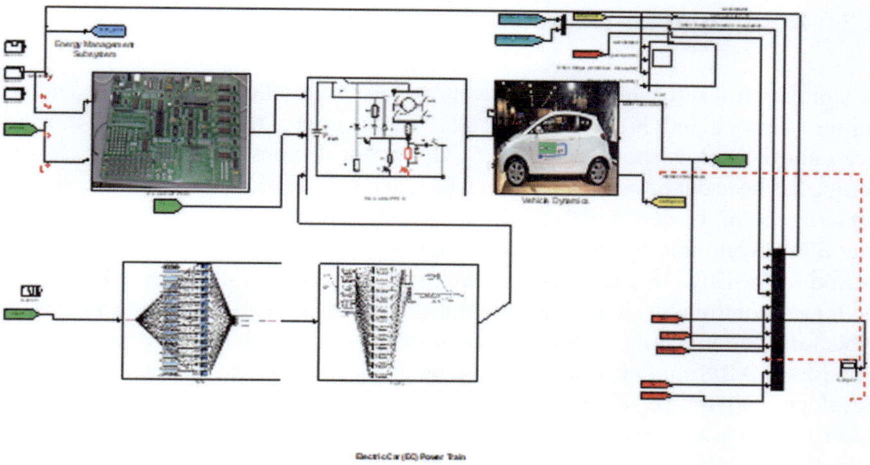

Fig. 8.4 MATLAB/Simulink model

current control comes next. Controlling the amount of current that flows to the brake resistor is how the IGBT V2 is used to manage the appropriate torque. Because the field current has previously been maintained, it is now possible to regulate the torque.

8.2.2 Simulation Model Development

For conducting study on the operation of resistive braking, a mathematical model [1, 2, 10], of a series motor and a FODC is utilized to create a simulation model, which can be shown in Figs. 8.4.

Using the MATLAB/Simulink software, a simulation model of a FQDC with a PID controller and an expert system controller is built up. There are two different models evaluated. A DC series motor with a capacity of 1 kW and an electric vehicle with a DC drive. The model is created by solving the linear differential equation of a DC series motor, and it is then represented in a physical-based model by making use of MATLAB/Simulink. First, a 0.6 kW DC series motor model is simulated and controlled using a four-quadrant chopper controlled by an expert system and PID, while LUT serves as the reference for the field current. With the help of the MATLAB/Simulink software, a simulation model of a FQDC, a PID controller, and an expert system controller are created. Both of these models are evaluated. A low-power kilowatt direct current series motor and a DC-drive electric vehicle.

8.3 Result and Discussion

A simulation result with lower kilowatt motor speed (shown in blue) and motor torque (shown in red) is shown in Fig. 8.5. The outcome of the simulation, which has been normalized, is shown in Fig. 8.5. This result demonstrates that the excessive torque has been cut off, stretched for a while, and it still follows the reference torque, up to the point when it is no longer manageable. The expert system and its control algorithm were able to successfully avoid the excessive torque and lengthen the period for resistive braking mode before the system entered a zone in which it was no longer controllable. This was accomplished by successfully avoiding a region in which the system could no longer be controlled. Figure 8.6 displays the outcomes of a MATLAB/Simulink simulation run on an electric vehicle operated by a newly developed FQDC. To successfully propel an electric vehicle with a total weight of 1325 kg using a series DC motor with a 35 kW motor that runs at a maximum power of 22 kW was tested. Figure 8.6 displays the results of a simulation run in MATLAB and Simulink that looked at the relationship between speed and torque during resistive braking mode chopper operation. In this mode, the speed of the vehicle drops quite significantly, as can be seen in the image. The energy that the motor's kinetic motion produces is converted into heat and then lost as heat. When the armature current (I_a) goes into the negative direction, the resultant torque in Eq. (8.3) [1, 2, 11], is negative. This is what causes the braking action to occur. Figure 8.6 displays, in light green, the torque that is generated by the electric car's motor while the vehicle is doing resistive braking. It is possible to apply resistive brakes up to a specific motor speed before the motor loses its ability to be controlled. The mechanical brake, which is activated by the brake pad [9], will begin to function once the electrical braking is uncontrollable or no longer necessary. When the motor vehicle reaches this speed, the mechanical brake will engage and take over stopping the vehicle. With the help of an expert system that is based on an AI controller, it is possible to manage the motor torque (in blue) as well as the produced voltage, which enables the resistive braking time to be further extended and controlled. In Fig. 8.6, the motor torque controlled with an AI (expert system) controller does not exceed the reference torque, in contrast to the motor that is not controlled with AI. This can be seen by comparing the blue portion of the motor torque to the reference torque. When compared to the variation that does not have AI assistance (shown in black), the slowdown of the motor speed that is regulated by AI (shown in red) is similarly more linear. Because the electrical braking can be further extended to even at lower speed, this will result in less time spent using the brake pads, which will result in the brake pads lasting longer. This will make driving an electric vehicle more pleasant for the driver as well as any passengers in the vehicle.

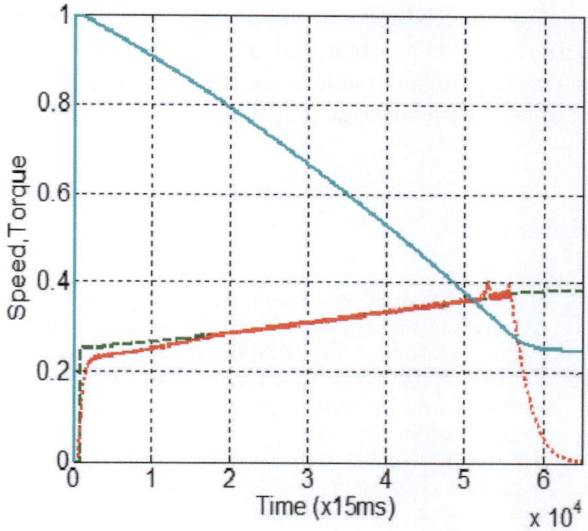

Fig. 8.5 Simulation result testing with lower kW motor

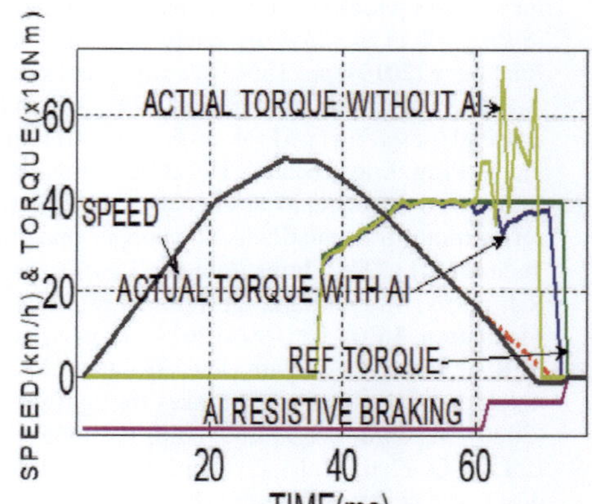

Fig. 8.6 Simulation result testing for DC drive EV model

8.4 Conclusion

This paper discusses the expert system and direct torque control to compensate the effect of excessive torque at the end of the resistive braking mode. From the obtained simulation result, it reveals both the expert system and direct torque control were effective in resolving the issue with excessive torque at the end of the resistive braking mode. It allowed for an extension of the amount of time spent using electrical brakes which is resistive braking mode without experiencing excessive braking torque. This not only makes the driver feel more at ease when operating the EV especially while braking, but it also provides an economic advantage since the brake pad may be used for a much longer period of time before it has to be changed. In conclusion, the performance of an optimized DC drive system that is coupled with an AI controller in resistive braking mode is equivalent to that of an AC drive. As a result, this kind of drive is particularly appropriate for use in applications that include low cost EVs.

References

1. S. Arof, E. Noorsal, P.A. Mawby, H. Arof, Series motor four quadrants drive DC chopper for DC drive electric vehicle part7: resistive braking, in *International Scientific Forum (ISF2019)* (Melaka, Malaysia, 2020). eISBN: 978-967-10842-1-2
2. S. Arof, J.A. Jalil, N.H. Kamaruddin, N.M. Yaakop, P.A. Mawby, H. Arof, Series motor four quadrants drive DC chopper part2: driving and reverse with direct current control. Int. Conf. Power Electron. 775–780. https://doi.org/10.1109/PECON2016.7951663
3. A. Fayad, H. Ibrahim, A. Ilinca, S. Sattarpanah Karganroudi, M. Issa, Energy recovering using regenerative braking in diesel–electric passenger trains: economical and technical analysis of fuel savings and GHG emission reductions. Energies **15**, 37 (2022). https://doi.org/10.3390/en15010037.
4. S. Arof, N.H.N. Diyanah, N.M. Noor, M. Rosyidi, P.A. Mawby, H. Arof, Genetics algorithm for setting up look up table for parallel mode of new series motor four quadrants DC chopper, in *Progress in Engineering Technology: Automotive, Energy Generation, Quality Control and Efficiency* (2019), pp. 155–167. https://doi.org/10.1007/978-3-030-28505-0_12
5. A.R.A. Saharul Arof, M. Shauqee, N.H.N. Rosyidi, P.M. Diyanah, H. Arof, E. Noorsal, Gradient descend for setting up a look-up table of series motor four quadrants drive DC chopper in parallel mode. Prog. Eng. Technol. III (2021). https://doi.org/10.1007/978-3-030-67750-3_16
6. S. Arof, A. Shauqee, M.K. Zaman, N.H.N. Diyanah, P. Mawby, H. Arof, E. Noorsal, Study on series motor four quadrants dc chopper operation controlled by an expert system. Prog. Eng. Technol. III (2021). https://doi.org/10.1007/978-3-030-67750-3_14
7. R.J. Farris, M. Goldfarb, Design of a multidisc electromechanical brake. IEEE/ASME Trans. Mechatron. **16**(6), 985–993 (2011). https://doi.org/10.1109/TMECH.2010.2064332
8. M.R.A. Putra, M. Nizam, D.P. Tjahjana, M. Aziz, A.R. Prabowo, Application of multiple unipolar axial eddy current brakes for lightweight electric vehicle braking. Appl. Sci. **10**, 4659 (2020). https://doi.org/10.3390/app1013465
9. L. Luo, Q. Zhai, W. Li, C. Qian, H. Liu, Research on an integrated electromagnetic auxiliary disc brake device for motor vehicle). IEEJ Trans. Electr. Electron. Eng. (2016). https://doi.org/10.1002/tee.22395
10. H. Shamieh, R. Sedaghati, Multi-objective design optimization and control of magnetorheological fluid brakes for automotive applications (IOP publishing, 2017). https://doi.org/10.1088/1361-665X/aa9452
11. P. Fajri, S. Heydari, N. Lotfi, Optimum low speed control of regenerative braking for electric vehicles, in *2017 IEEE 6th International Conference on Renewable Energy Research and Applications (ICRERA)* (San Diego, CA, USA, 2017), pp. 875–879. https://doi.org/10.1109/ICRERA.2017.8191185

Chapter 9
Sliding Mode in Direct Current Control of Driving Mode of Direct Current Drive Electric Vehicles

Norhalimatul Sadiah Kamaruddin, Halina Hassan, Amir Shauqee Abdul Rahman, M. Rezal, Saharul Arof, Philip Mawby, Hamzah Arof, and Emilia Noorsal

Abstract Direct current (DC) series motors have the highest starting torque of similar kilowatt motors. For a DC drive electric car with a series motor, the conventional close loop speed based control causes not only jerk but tire slip during startup. To overcome this problem, in this paper direct current control is applied to a four quadrants DC chopper (FQDC) to solve the startup problem. A sliding mode controller is used as close loop feedback to control the fluctuated series motor rising current during start up. The controller offers robust and disturbance rejection advantages.

N. S. Kamaruddin (✉) · H. Hassan · A. S. A. Rahman · M. Rezal · S. Arof
Universiti Kuala Lumpur, Malaysian Spanish Institute Kulim Hi-Tech Park, 09000 Kulim, Kedah, Malaysia
e-mail: norhalimatul@unikl.edu.my

H. Hassan
e-mail: halina@unikl.edu.my

A. S. A. Rahman
e-mail: amirshauqee@unikl.edu.my

M. Rezal
e-mail: mrezal@unikl.edu.my

S. Arof
e-mail: saharul@unikl.edu.my

S. Arof · P. Mawby
University of Warwick School of Engineering, Coventry CV47AL, UK
e-mail: P.A.Mawby@warwick.ac.uk

H. Arof
Engineering Department, Universiti Malaya, Jalan Universiti, 53600 Kuala Lumpur, Malaysia
e-mail: ahamzah@um.edu.my

E. Noorsal
Electrical Engineering Studies, College of Engineering, Universiti Teknologi MARA Cawangan Pulau Pinang, 13500 Permatang Pauh, Pulau Pinang, Malaysia
e-mail: emilia.noorsal@uitm.edu.my

T. A. A. Razak et al. (eds.), *IT Application for Sustainable Living 2*,
SpringerBriefs in Applied Sciences and Technology,
https://doi.org/10.1007/978-3-031-93198-7_9

MATLAB/Simulink simulation software is used to simulate and test the FQDC during startup which is running in driving mode chopper operation. The controlled sliding mode result shows good performance beyond acceptable.

Keywords Direct current (DC) motor · Direct current control · Four-quadrant direct current chopper · Sliding mode controller

9.1 Introduction

Reducing global hydrocarbon emissions [1] through the use of efficient electric vehicles (EVs) for transportation is a viable option. However, the high cost of existing alternating current (AC) drive EVs makes them inaccessible to many people, particularly in developing nations. Therefore, studies are needed to discover cost-effective EV alternatives, such as direct current (DC) drive EVs [1]. Research on DC drive EVs has introduced a new four-quadrant DC chopper (FQDC) for series motors [1], which offers more operational flexibility. Attempts have been made to explore separately excited [2] motors as an alternative to series motors, with some adjustments to the FQDC. A simulation model of the DC EV and FQDC [3] has been established and used for various purposes, including EV design requirements [4], comparative studies [5], and detailed investigations of chopper operations [6] concerning current, voltage, and torque [7]. PID controllers tuned using pole placement [8] and fuzzy controllers [9] have been employed for closed-loop feedback direct current control [10]. This paper discusses a sliding mode closed-loop feedback controller for regulating the current in the direct current control of FQDC during the driving mode of a DC series motor.

9.1.1 Four-Quadrant Drive DC Chopper

The FQDC, as shown in Fig. 9.1, has seven modes of operation [1–6]. Figure 9.1 illustrates the FQDC circuit in driving mode for closed-loop feedback control of a DC series motor during startup and the flow of current in driving mode is also addressed in [10].

During startup, jerk caused by excessive starting torque is undesirable. Direct torque control and direct current control, which are discussed in detail in [10], help reduce jerk and high starting torque. Fuzzy logic and digital PID controllers have been tested to control motor speed and current using direct current control [8, 9]. For the PID controller, the pole placement method [8] is used to adjust the closed-loop PID controller to enhance its performance. Meanwhile, the fuzzy logic controller [9] is employed to improve both control performance and system stability.

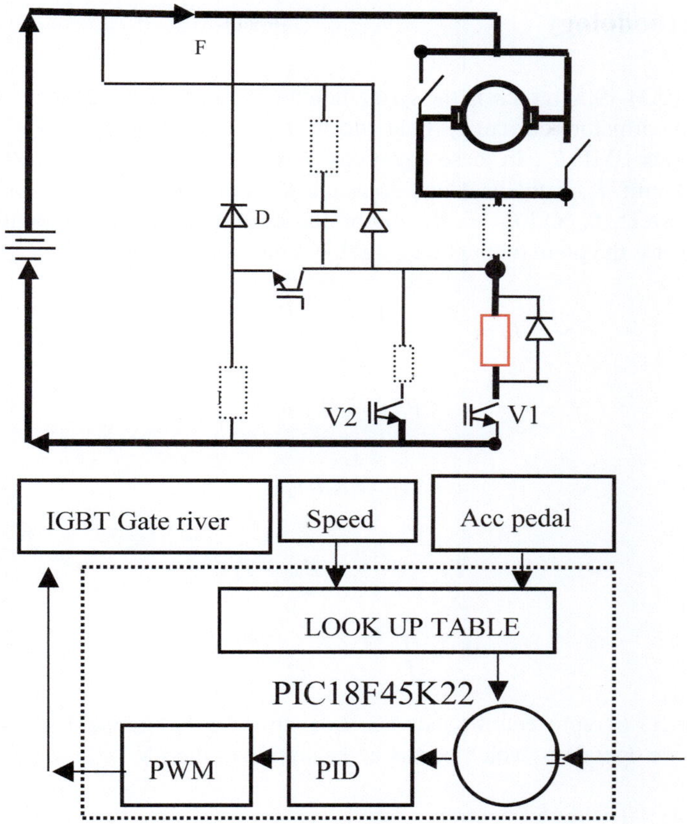

Fig. 9.1 Driving mode of the FQDC

9.1.2 Sliding Mode Control

A sliding mode controller (SMC) [11, 12] is required to address practical issues related to disturbance rejection and setpoint tracking. SMC employs continuous non-linear feedback, where the control input rapidly switches between two or more limits when both the error and its rate become zero. As a result, the system state crosses and recrosses the sliding surface or switching surface before settling on it. The SMC emphasizes the sliding motion, with the control strategy determining the path that the sliding surface will follow.

9.2 Methodology

Equations (9.1)–(9.5) represent the series motor equations, showing the relationships between the armature current I_a, field current I_f, voltage supply V_{dc}, PWM output K_{drv} and back EMF E_g. In these equations, I_a is the armature current, B_{emf} is the motor back emf, K_v is the back emf constant, K_t is the torque motor constant, I_f is the field current, R_a and R_f are the motor coil resistances, ω is the angular speed, K_{drv} represents the pwm output and T_d *is* the motor torque [1–3].

$$I_a = \frac{V_{batt} - I_a(R_a + R_f) - B_{emf}}{L_a + L_f} \tag{9.1}$$

$$B_{emf} = K_v I_f \omega \tag{9.2}$$

$$T_d = K_t I_f I_a \tag{9.3}$$

$$T_d = J\frac{d\omega}{dt} + B_w + T_L \tag{9.4}$$

$$W = \int \frac{T_d - (T_L + B_w)}{J} \tag{9.5}$$

Equation (9.6) represents the state-space equation for the chopper leg [1–3] during driving mode operation, which occurs at the startup of the EV.

$$\frac{d}{dt}\begin{bmatrix} I_a \\ I_f \\ V_c \end{bmatrix} = \begin{bmatrix} \frac{-(R_a+R_c)}{(L_a)} & \frac{R_C}{(L_a)} & \frac{1}{(L_a)} \\ \frac{R_c}{(L_f)} & \frac{-(R_f+R_c)}{(L_f)} & \frac{1}{(L_f)} \\ \frac{-1}{(C)} & \frac{1}{(C)} & 0 \end{bmatrix}\begin{bmatrix} I_a \\ I_f \\ V_c \end{bmatrix} + \begin{bmatrix} 0 & 0 & 0 \\ \frac{1}{L_f} & \frac{-1}{L_f} & 0 \\ 0 & 0 & 0 \end{bmatrix}\begin{bmatrix} V_{dc}K_{drv} \\ E_g \\ 0 \end{bmatrix} \tag{9.6}$$

For controlling the four quadrant DC chopper (FQDC) in driving mode using the direct current control technique [10], Eq. (9.6) is transformed into a transfer function where the motor current is the controlled variable and the input terminal voltage, modulated by PWM, is the manipulating variable.

$$\begin{aligned} \frac{\omega_n^2}{s^2 + 2\zeta\omega_n s + \omega_n^2} &= \frac{J_m^S + B_m}{L_{a+f}\cdot J_m \cdot S^2 + (R_{a+f}\cdot J_m + B_m.L_{a+f})S + (K_T \cdot K_e)} \\ &= \frac{I_a(S)}{V_{dc}(S)} \end{aligned} \tag{9.7}$$

In the time domain, the above equation can be expressed as:

$$L_{a+f}\cdot J_m \cdot \ddot{\omega}(t) + (R_{a+f}\cdot J_m + B_m.L_{a+f})\dot{\omega}(t)$$

Table 9.1 DC motor parameters

$R_a = 0.6\,\Omega$	$J = 0.0167$ kg m^2/s^2
$K_e = 0.8$ rad/s	$K_t = 0.8$ Nm/A
$L_a = 0.012$H	$b = 0.0167$

$$+\left(K_T \cdot K_e + R_{a+f} \cdot B_m\right)\omega(t) = \left(J_m^S + B_m\right)V_{dc}(t) \tag{9.8}$$

where the parameters listed in Table 9.1 are assigned as follows:

Replacing the parameters to the second order transfer function will result in Eq. (9.9):

$$\frac{1}{s^2 + 4.5318s + 3} \tag{9.9}$$

In the state space domain, the denominator of transfer function can be written as Eq. (9.10):

$$\ddot{\omega}(t) + 4.5318\dot{\omega}(t) + 3\omega(t) = V_{dc}(t) \tag{9.10}$$

Replace X_1 and u as Eq. (9.11):

$$X_1 = \omega(t) \text{ and } u = V_{dc}(t) \tag{9.11}$$

Then, the system can be converted in the following canonical form [11, 12].

$$\dot{X}_1 = X_2 \tag{9.12}$$

$$\dot{X}_2 = -3X_1 - 4.5318X_2 + U \tag{9.13}$$

$$Y = X_1 \tag{9.14}$$

Now select the sliding surface

$$\sigma = c(r - X_1) + X_2 \tag{9.15}$$

where c is constant of sliding matrix $c\ \sigma\epsilon\ R$^(m*n) such that $c < 0$. The total control law is given by

$$U = U_1 + U_{nl} \tag{9.16}$$

$$U = -3X_1 + (C - 4.5318)X_2 + K\frac{\sigma}{\sigma + \delta} \tag{9.17}$$

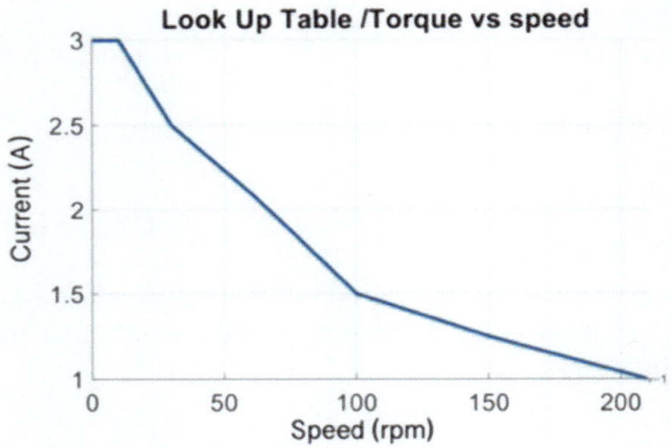

Fig. 9.2 The motor current reference in look up table for direct current control for speed versus current

where $K > 0$ is chosen to be sufficiently large. The larger the value of K, the faster the trajectory converges to the sliding surface. The value of K can be determined through trial and error or by using optimization tools such as particle swarm optimization (PSO), genetic algorithms (GA) [13], or gradient descent (GD) [14].

The simulation model for the DC series motor was developed using MATLAB/Simulink software and is based on Eqs. (9.1)–(9.6). It includes a sliding mode controller for testing the controller's performance in controlling the system.

9.3 Results

The lookup table (LUT), which is essential for direct current control [8–10], is shown in Fig. 9.2. The LUT provides the input reference for the expected motor current at each motor speed of the series motor [8–10] to the sliding mode controller. The sliding mode controller then outputs the PWM signal to control the IGBT firing of V_1. Firing the IGBT V_1 results in an increase in armature current, as shown in Fig. 9.3. The field current is equivalent to the armature current because it is connected in series during the driving mode, which is operational during startup.

From the graph in Fig. 9.3, the output current (red) closely follows the reference current (blue) pattern provided by the LUT. Although there is an overshoot shown in Fig. 9.3, it is considered minimal since the current responds quickly to control adjustments.

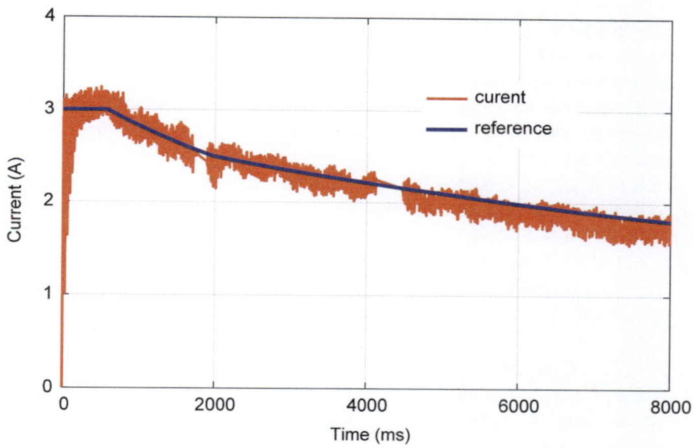

Fig. 9.3 Close loop current control performance using sliding mode controller

9.4 Conclusion

The SMC effectively controls the armature current of the FQDC in the driving mode of a series motor for a DC drive electric car, utilizing the LUT during EV startup. The control performance is considered to be excellent and exceeds acceptable standards.

References

1. M Muratori E Elgqvist D Cutler J Eichman S Salisbury Z Fuller J Smart 2019 Technology solutions to mitigate electricity cost for electric vehicle DC fast charging Appl. Energy 242 415 423
2. J Santiago De H Bernhoff B Ekergård S Eriksson S Ferhatovic R Waters M Leijon 2011 Electrical motor drivelines in commercial all-electric vehicles: a review IEEE Trans. Veh. Technol. 61 2 475 484
3. S. Arof, H. Hassan, M. Rosyidi, P.A. Mawby, H. Arof, Implementation of series motor four quadrants drive DC CHOPPER for DC drive electric car and LRT via simulation model. J. Appl. Environ. Biol. Sci. **7**(3S), 73–82 (2017)
4. M. Chae, Y. Hyun, K. Yi, K. Nam, Dynamic handling characteristics control of an in-wheel-motor driven electric vehicle based on multiple sliding mode control approach. IEEE Access **7**, 132448–132458 (2019)
5. K Kakouche T Rekioua S Mezani A Oubelaid D Rekioua V Blazek L Prokop S Misak M Bajaj SSM Ghoneim 2022 Model predictive direct torque control and fuzzy logic energy management for multi power source electric vehicles Sensor 5669 1 22
6. L He W Ye Z He K Song Q Shi 2020 A combining sliding mode control approach for electric motor anti-lock braking system of battery electric vehicle Control. Eng. Pract. 102 104520
7. M. Aktas, K. Awaili, M. Ehsani, A. Arisoy, Direct torque control versus indirect field-oriented control of induction motors for electric vehicle applications. Eng. Sci. Technol. Int. J. 2215–0986 (2020)

8. H Wei N Zhang J Liang Q Ai W Zhao T Huang Y Zhang 2022 Deep reinforcement learning based direct torque control strategy for distributed drive electric vehicles considering active safety and energy saving performance Energy 238 121725
9. A Oubelaid H Alharbi AS Humayd N Taib T Rekioua SSM Ghoneim 2022 Fuzzy-energy-management-based intelligent direct torque control for a battery—supercapacitor electric vehicle Sustainability. 14 8407
10. AK Singh KK Prabhakar CU Reddy P Kumar 2020 A lookup table-based reference flux linkage selection of direct torque control induction motor drive for electric vehicle applications SAE Int. J. Electrified Veh. 9 1 61 90
11. A. Ma'arif, A. Çakan, Simulation and Arduino hardware implementation of DC motor control using sliding mode controller. J. Robot. Control. (JRC) **2**(6) (2021)
12. C.U. Maheswararao, Y.S.K. Babu, K. Amaresh, Sliding mode speed control of a DC motor, in *Proceeding 2011 International Conference on Communication Systems and Network Technologies* (2011), pp. 387–391
13. Y Liu Y Wei C Wang H Wu 2024 Trajectory optimization for adaptive deformed wheels to overcome steps using an improved hybrid genetic algorithm and an adaptive particle swarm optimization Mathematics 12 13 2077
14. AAA Rasheed AS Al-Araji MN Abdullah 2022 Static and dynamic path planning algorithms design for a wheeled mobile robot based on a hybrid technique Int. J. Intell. Eng. Syst. 15 4 167 181

Chapter 10
Series Motor Four Quadrants DC Chopper for Economical Electric Vehicle: Modeling and Simulation of the Braking Operation

Norhalimatul Sadiah Kamaruddin, Amir Shauqee Abdul Rahman, Muhammad Sazali Said, Saharul Arof, Philip Mawby, H. Arof, and Emilia Noorsal

Abstract Simulation models are crucial to study systems that are essential in technical applications. However, such models require time to build and they are costly. The established simulation model needs to reveal a fast convergence and must run a long time without a failure due to a shortage of memory. This paper presents a mathematical model of a four-quadrant DC chopper (FQDC) for electric cars (EC) in regenerative and resistive braking modes. The established simulation model offers fast convergence and long simulation duration without failure. Physics based, transfer

N. S. Kamaruddin (✉) · A. S. A. Rahman · M. S. Said · S. Arof
Universiti Kuala Lumpur, Malaysian Spanish Institute Kulim Hi-Tech Park, 09000 Kulim, Kedah, Malaysia
e-mail: norhalimatul@unikl.edu.my

A. S. A. Rahman
e-mail: amirshauqee@unikl.edu.my

M. S. Said
e-mail: msazali@unikl.edu.my

S. Arof
e-mail: saharul@unikl.edu.my

S. Arof · P. Mawby
University of Warwick School of Engineering, Coventry CV47AL, UK
e-mail: P.A.Mawby@warwick.ac.uk

H. Arof
Engineering Department, Universiti Malaya, Jalan Universiti, 50603 Kuala Lumpur, Malaysia
e-mail: ahamzah@um.edu.my

E. Noorsal
Electrical Engineering Studies, College of Engineering, Universiti Teknologi MARA Cawangan Pulau Pinang, 13500 Permatang Pauh Campus, Permatang Pauh, Pulau Pinang, Malaysia
e-mail: emilia.uitem@edu.my

T. A. A. Razak et al. (eds.), *IT Application for Sustainable Living 2*,
SpringerBriefs in Applied Sciences and Technology,
https://doi.org/10.1007/978-3-031-93198-7_10

function and state space models are adopted to simplify and solve linear differential equations to establish the simulation model. To verify the mathematical model, the experimental results will be compared. The mathematical model is tested using the MATLAB/Simulink software, and the simulation and experimental results show that the model can represent the braking modes of the proposed FQDC chopper operations.

Keywords Simulation model · Four-quadrant DC chopper (FQDC) · Electric cars (EC) · Mathematical model · Sliding mode

10.1 Introduction

When doing research on potentially dangerous systems that are difficult, time-consuming, and expensive to build, and difficult to get test subjects (human or animal), simulation models are very necessary [1]. The mathematical model of the electric vehicle (EV) was built for a variety of purposes, one of which is to carry out research on various aspects of the power train [2–4]. In the process of doing study and investigation into the power train that is driven by the power converter [2–4], mathematical modeling may also be applied. Steady state model (SS), the linear differential equation (LDE), the physical-based model (PBM), and the transfer function model (TFM) are the most common types of mathematical models used to establish simulation model. The ORNL [5–12], designed in 2009, a DC brushed motor with 55 kW power output, 92% efficiency, and 13 V operating voltage, which boosted the interest in DC drive EC research. A newly proposed four quadrants DC chopper (FQDC) with driving, field weakening, parallel mode, regenerative braking, and resistive braking operations to replace conventional H-bridge choppers [7–11]. A simulation model based on drag and drop icon was established to further detail the investigation on electric car design requirement, or each of the chopper operations and on the specific pattern of motor voltage, current, torque, speed of the series motor, and FQDC running for DC drive EC application [7–11]. While testing the DC drive EV performance for a new European driving cycle (NEDC) test using the simulation model developed using the MATLAB/Simulink drag and drop icon, the software happens running slow and eventually the simulation stop due to out of memory problem. To solve this problem, a simulation model based on LDE converted to a physical based model is required. The model can also run faster and longer simulation time is possible. A mathematical model of the FQDC chopper is needed to study transient response, system stability, and the close loop feedback controller design. The mathematical model is also used to develop a system representation using a numerical representation such as using Taylor series [12] for faster execution and longer simulation time. Mathematical models in braking mode are needed for NEDC driving cycle test simulation because they allow faster simulation and longer simulation time. This paper continues the discussion on discussing braking mode mathematical modeling.

10.2 Literature Review

10.2.1 Regenerative Braking

The detail of regenerative braking has been discussed in [6]. In this paper, we cover the mathematical equation that has not been covered in [6]. The electric illustrative circuit of regenerative braking for voltage development using resonator circuit is shown in Fig. 10.1. In the following mathematical equations, k_{drv} is the chopper duty cycle for V1, kf_w is the chopper duty cycle for V2, and k_{rgb} is the chopper duty cycle for V3.

During voltage development using the resonator circuit and the commutation stage, the electrical illustrative circuit is as in Fig. 10.2 and the mathematical equations are:

$$V_{\text{dc}}K_{\text{drv}}K_{\text{rgb}} = V_c + I_f R_c + I_f R_f + L_f \frac{\text{dif}}{\text{dt}} \quad (10.1)$$

$$\frac{\text{dif}}{\text{dt}} = \frac{V_{\text{dc}}K_{\text{drv}}K_{\text{rgb}} - V_c - I_f\left(R_c + R_f\right)}{L_f} \quad (10.2)$$

State Space

$$\begin{bmatrix} \frac{-(R_c+R_f)}{L_f} & \frac{-1}{L_f} \\ \frac{I_f}{C} & 0 \end{bmatrix} \begin{bmatrix} I_f \\ V_c \end{bmatrix} \begin{bmatrix} \frac{1}{L_f} & 0 \end{bmatrix} \begin{bmatrix} V_{\text{dc}}K_{\text{drv}}K_{\text{rgb}} \\ 0 \end{bmatrix} \quad (10.3)$$

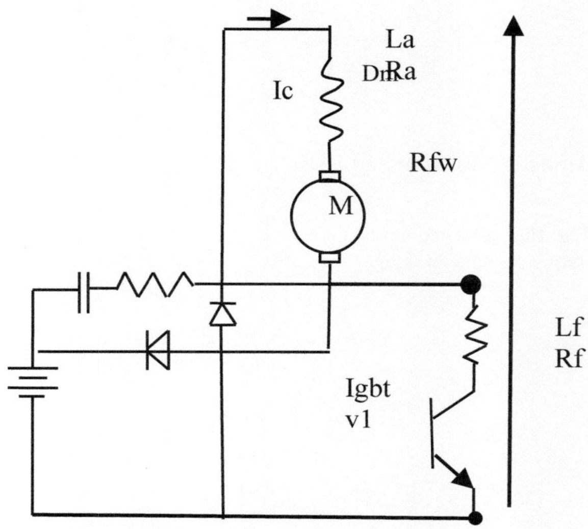

Fig. 10.1 Flow of current during regenerative mode in voltage development and commutation state

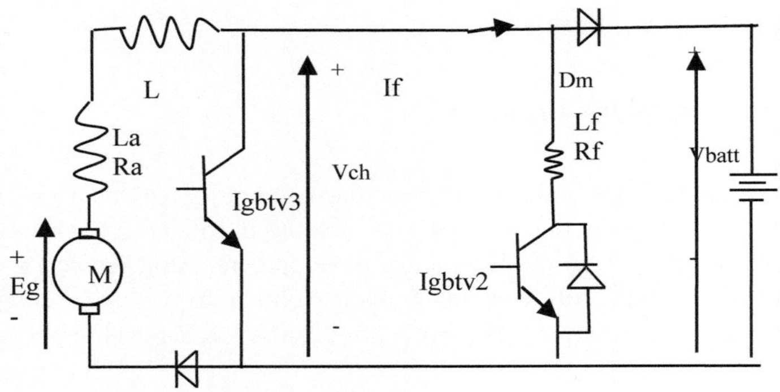

Fig. 10.2 Flow of current during regenerative mode in braking state

$$\begin{bmatrix} \dot{V}_c \\ I_f \end{bmatrix} = \begin{bmatrix} 0 & \frac{-1}{C} \\ \frac{-1}{L} & \frac{-R}{L} \end{bmatrix} \begin{bmatrix} V_c \\ I_f \end{bmatrix} + \begin{bmatrix} 0 \\ \frac{1}{L} \end{bmatrix} \begin{bmatrix} V_{dc} \end{bmatrix} \tag{10.4}$$

During braking stage and the commutation stage the electrical illustrative circuit is as in Fig. 10.3 and the mathematical equations in linear differential equation and transfer function are:

LDE

$$E_g K_{rgb} = I_{short} R_a + L_a \frac{dI_{short}}{dt} + I_{short} R_{short} \tag{10.5}$$

$$\frac{di_{short}}{dt} = \frac{E_g K_{rgb} - I_{short}(R_a + R_{short})}{(L_a)} \tag{10.6}$$

$$\left[-\left(\frac{R_a + R_{short}}{L_a} \right) \right] [I_{short}] \left[\frac{1}{(L_a)} \right] \left[E_g K_{rgb} \right] \tag{10.7}$$

Transfer Function and LDE

Fig. 10.3 Commutation during resistive braking

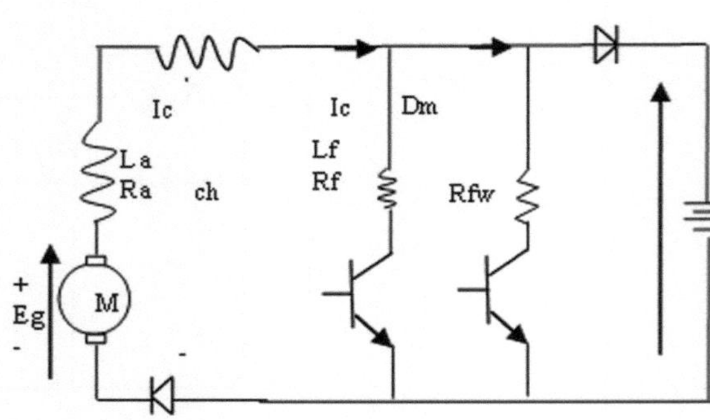

$$(L_a)\frac{di_{\text{short}}}{dt} = E_g K_{\text{rgb}} - I_{\text{short}}(R_a + R_{\text{short}}) \tag{10.8}$$

$$I_{\text{short}}(s) = \frac{E_g K_{\text{rgb}}(1-U)}{s(L_a + L_f) + (R_{\text{short}} + R_a)} \tag{10.9}$$

$$E_g - V_{\text{batt}} = I_{\text{batt}} R_a + L_a \frac{dI_{\text{batt}}}{dt} R_f + I_{\text{batt}} R_{\text{batt}} \tag{10.10}$$

$$\frac{dI_{\text{batt}}}{dt} = \frac{E_g - V_{\text{batt}} - I_{\text{batt}}(R_a + R_{\text{batt}})}{(L_a)} \tag{10.11}$$

10.2.2 Flow of Current During Resistive Braking

Details on resistive braking has been discussed in [5]. In this paper, we cover the mathematical aspects that has not been shown in [5]. The circuit representation of resistive braking is shown in Fig. 10.3.

In Fig. 10.3, *V1* is the main IGBT, R_{fw} is the braking series resistor that used to brake the vehicle using IGBT *V2,* and k_{drv} chopper duty cycle *V1*, k_{fw} chopper duty cycle *V2*.

The equations that relate those parameters in linear differential equations representation are:

Linear differential Equation

$$E_g - V_{\text{batt}} = I_{\text{batt}} R_a + L_a \frac{dI_{\text{batt}}}{dt} + I_{\text{batt}} R_{\text{batt}} \tag{10.12}$$

$$\frac{dI_{batt}}{dt} = \frac{E_g - V_{\text{batt}} - I_{\text{batt}}(R_a + R_{\text{batt}})}{(L_a)} \tag{10.13}$$

State Space

$$\left[-\left(\frac{R_a + R_{\text{batt}}}{L_a}\right)\right][I_{\text{batt}}]\left[\frac{1}{(L_a)}\right][E_g - V_{\text{batt}}] \tag{10.14}$$

Transfer Function

$$s(L_a)I_{\text{batt}}(s) + I_{\text{batt}}(s)(R_{\text{batt}} + R_a) = (E_g - V_{\text{batt}})(s) \tag{10.15}$$

$$I_{\text{batt}}(s) = \frac{E_g - V_{\text{batt}}}{sL_a + (R_{\text{batt}} + R_a)} \tag{10.16}$$

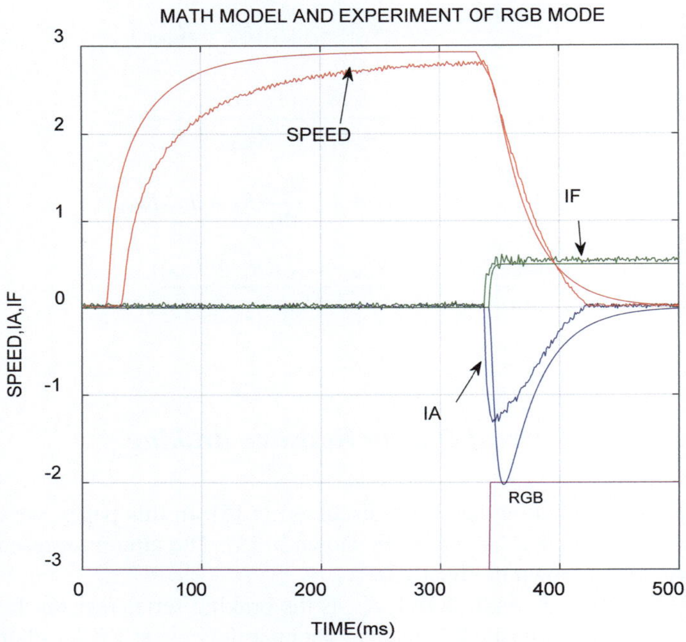

Fig. 10.4 Regenerative braking mode

10.3 Results

The experimental results using the Kaptoris software (normalized) in [2, 3], are combined with the established mathematical model and the results are shown in Figs. 10.4 and 10.5. The motor speed (red) was initially increased until saturation (representing FQDC driving and field weakening modes). The motor speed (in red in Figs. 10.4 and 10.5), armature current (blue in Figs. 10.4 and 10.5), field current (green in Figs. 10.4 and 10.5), as shown in Figs. 10.4 and 10.5. The mathematical simulation model for regenerative and resistive braking is as discussed in [2, 3]. IGBT V3 shorts the armature during regenerative braking (Fig. 10.4). The armature current rose oppositely, creating torque to oppose the motor motion. The generated current flows via short circuit, field, and armature resistor [2, 3] to create the braking effect. The generated voltage at the armature can be used for charging the battery. In resistive braking, the braking resistor dissipated the generated power in voltage and current. The same current also flew to the motor field to generate voltage, to charge battery and to generate braking effect (Fig. 10.5).

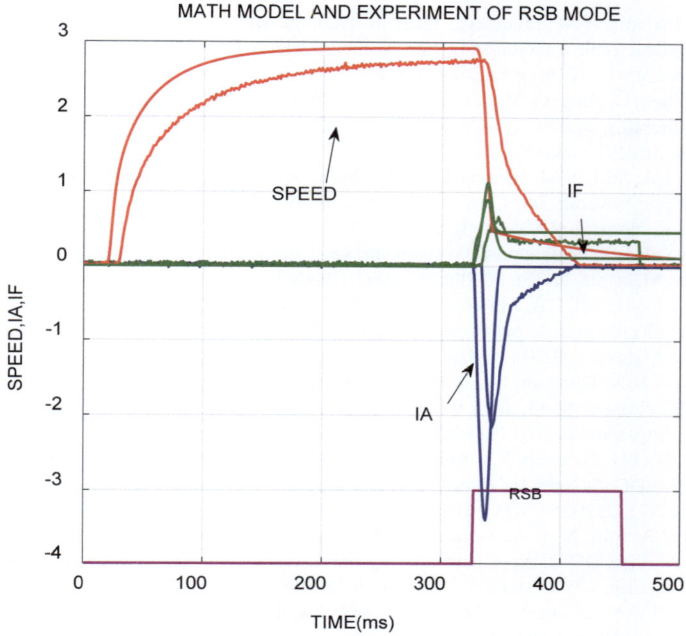

Fig. 10.5 Regenerative-Resistive braking mode

10.4 Conclusion

This paper discusses three mathematical modeling which is physics based model, transfer function and state space equations. From the results, it can be concluded that the mathematical models can verify and represent the actual hardware allowing that the study can be done fast, safe, and reliable. The mathematical model when converted to numerical representation can be used for simulating EV for NEDC test which requires longer simulation time when the system is tested with the FQDC, controller with AI algorithms and higher kilowatt motor.

References

1. A.J. Kucharski, T.W. Russell, C. Diamond, Y. Liu, J. Edmunds, S. Funk, R.M. Eggo, Early dynamics of transmission and control of COVID-19: a mathematical modelling study. Lancet. Infect. Dis (2020). https://doi.org/10.1016/S1473-3099(20)30144-4
2. S. Abdelrahman, K.S. Algarny, M.Z. Youssef, A Novel platform for powertrain modeling of electric cars with experimental validation using real-time hardware in the loop (HIL): a case study of GM second generation Chevrolet volt. IEEE Trans. Power Electron. **33**(11), 9762–9771 (2018). https://doi.org/10.1109/TPEL.2018.2793818

3. B.S.K. Patnam, N.M. Pindoriya, Demand response in consumer-centric electricity market: mathematical models and optimization problems. Electr. Power Syst. Res. **193**, 106923 (2021). https://doi.org/10.1016/j.epsr.2020.106923
4. Z.J.M. Shen, B. Feng, C. Mao, L. Ran, Optimization models for electric vehicle service operations: a literature review. Transp. Res. Part B: Methodol. **128**, 462–477 (2019). https://doi.org/10.1016/j.trb.2019.08.006
5. S. Arof, J.A. Jalil, N.M. Yaakop, P.A. Mawby, H. Arof, Series motor four quadrants drive DC chopper part1:overall. Int. Conf. Power Electron. (2014)
6. S. Arof, E. Noorsal, P.A. Mawby, H. Arof, Series motor four quadrants drive DC chopper for DC drive electric vehicle part7: resistive braking, in *International Scientific Forum (ISF2019)* (Melaka, Malaysia, 2020). eISBN: 978-967-10842-1-2
7. S. Arof, E. Noorsal, P.A. Mawby, H. Arof, Series motor four quadrants drive DC chopper for DC drive electric vehicle part6: regenerative braking. International Scientific Forum (ISF2019) (Melaka, Malaysia, 2020). eISBN: 978-967-10842-1-2
8. S. Arof, N.H.N. Diyanah, P.A. Mawby, H. Arof Emilia Noorsal, Series motor four quadrants drive DC chopper part3: field weakening mode. IOP Conf. Ser. Mater. Sci. Eng. **1127**(1), 012028. https://doi.org/10.1088/1757-899X/1127/1/012028
9. S. Arof, N.H.N. Diyanah, P.A. Mawby, H. Arof Emilia Noorsal, New series motor four quadrants drive DC chopper for economical EV, Part5: parallel mode. IOP Conf. Ser. Mater. Sci. Eng. **1127**(1), 012025 (2021). https://doi.org/10.1088/1757-899X/1127/1/012025
10. S. Arof, J.A. Jalil, N.H. Kamaruddin, N.M. Yaakop, P.A. Mawby, H. Arof, Series motor four quadrants drive DC chopper part2:driving and reverse with direct current control. Int. Conf. Power Electron. 775–780. https://doi.org/10.1109/PECON2016.7951663
11. S. Arof, N.H.N. Diyanah, N.M.N. Noor, Md. Radzi, J.A. Jalil, P.A. Mawby, H. Arof, Series motor four quadrants drive DC chopper: part 4: generator mode, in *Progress in Engineering Technology: Automotive, Energy Generation, Quality Control and Efficiency*, pp. 155–167. https://doi.org/10.1007/978-3-030-28505-0_12
12. S. Arof, E.D. Sukiman, N.H. Diyanah, N.M. Noor, Emilia Noorsal, P.A. Mawby, H. Arof, Discrete-time linear system of new series motor four-quadrant drive direct current chopper numerically represented by Taylor series, in *Progress in Engineering Technology II. Advanced Structured Materials*, vol. 131 (Springer, Cham, 2020). https://doi.org/10.1007/978-3-030-46036-5_10

Chapter 11
Gradient Descend Approach for the Setting up of a Look-Up Table in the Field Weakening Mode of a Four Quadrants DC Chopper for Electrical Vehicles

Norramlee Mohamed Noor, Saharul Arof, Abdul Malik Mohd Ali, Hamzah Arof, and Emilia Noorsal

Abstract The field weakening mode allows a DC series motor to operate at higher speeds. To avoid chopper failure in this mode of operation, the field current must be regulated according to the motor speed. If the chopper fails, it drains a large amount of current from the battery, but it produces little torque, causing the motor speed to decrease below the base speed, which is undesirable. The purpose of this study is to explore the creation of a look-up table (LUT) of speed versus field current in the field weakening mode of four quadrants DC (FQDC). During the field weakening mode, the LUT is required as an input reference for the close-loop feedback control of the field current. Gradient descent (GD) is used to help to create the LUT. The LUT is utilized when the FQDC is in field weakening mode, which causes the motor to

N. M. Noor (✉) · S. Arof · A. M. M. Ali
Universiti Kuala Lumpur, Malaysian Spanish Institute Kulim Hi-Tech Park, Kulim, Kedah, Malaysia
e-mail: noramlee@unikl.edu.my

S. Arof
e-mail: saharul@unikl.edu.my

A. M. M. Ali
e-mail: amalik@unikl.edu.my

S. Arof
University of Warwick School of Engineering, Coventry, UK

H. Arof
Engineering Department, Universiti Malaya, Jalan Universiti, Kuala Lumpur, Malaysia
e-mail: ahamzah@um.edu.my

E. Noorsal
Faculty of Electrical Engineering, Universiti Teknologi MARA Cawangan Pulau Pinang, Permatang Pauh, Pulau Pinang, Malaysia
e-mail: emilia.uitem@edu.my

T. A. A. Razak et al. (eds.), *IT Application for Sustainable Living 2*,
SpringerBriefs in Applied Sciences and Technology,
https://doi.org/10.1007/978-3-031-93198-7_11

run faster than its base speed. The FQDC and the resulting LUT are tested using the MATLAB/Simulink simulation model. When evaluated utilizing the suggested four quadrants DC chopper for EV applications, the simulation results demonstrate that the GD effectively found the best ideal field current to produce the highest torque.

Keywords Field weakening · Look-up table · Gradient descent · EV · FQDC

11.1 Introduction

Electric vehicles (EV), which are currently gaining popularity against conventional internal combustion engine-based transportation vehicles, are intended to overcome environmental pollution. The EV can be either AC or DC-powered. Depending on the AC or DC motor, field weakening is a phase in which the motor operates above its base speed region. In process optimization, tools such as gradient descent, genetic algorithms, and particle swarm optimization are used. Optimization is the process of finding the optimum value of target parameters by varying a control parameter. The output of the optimization process, or the final result, can be in the form of a look-up table (LUT), which stores information on control parameters such as speed versus motor current. The process can be done mostly offline or online using a simulation model. An electric vehicle with a DC motor is more cost-effective than one with an AC motor. This is due to the AC-driven vehicle's high cost and lack of appeal to residents of underdeveloped nations. Third-world nations must protect the environment from polluted emissions. A look-up table for field current can be built using an AI optimization technique, such as the genetic algorithm [1] or gradient descent [2], which improves performance in each specific four quadrants DC (FQDC) chopper operating mode. As a result, readings of the field current will be more precise. In this paper, the construction of a look-up table (LUT) of speed versus field current in the field weakening mode of FQDC electric vehicles is discussed. The LUT is required as an input reference for the closed-loop feedback regulation of the field current during the mode in which the field strength is reduced. The gradient descent algorithm is applied to assist in the development of the LUT. A convex optimization method called gradient descent (GD) looks for a local minimum. It is widely used and has numerous applications in a multitude of fields. Several modes of operation are necessary for electric vehicles powered by DC motors to operate properly. One of the modes of operation is called the field weakening mode, where the series DC motor runs above its base speed. GD can be used to optimize this mode by providing the optimum values of the field current at various speeds that can be stored in a look-up table.

11.1.1 A New Design of Four Quadrant Drive DC Chopper

The distribution of motor current via the primary IGBT V1 and IGBT V2 is shown in Fig. 11.1. When the IGBT V2 is turned on, current flows through the field weakening resistor instead of the field winding since it has a lower resistance. Field weakening operations where the motor torque is boosted to overcome load effects or when the EV tends to increase its speed have been discussed in detail in [3]. In order to discover the best torque by controlling the field current in accordance with the motor speed, the establishment of a LUT is discussed in this work. The best way to maintain torque performance is to construct a LUT using an expert system assistance and an optimization tool like the gradient descend algorithm [4]. This is because the field current and armature current are directly related to the motor torque but inversely proportional to speed and back EMF.

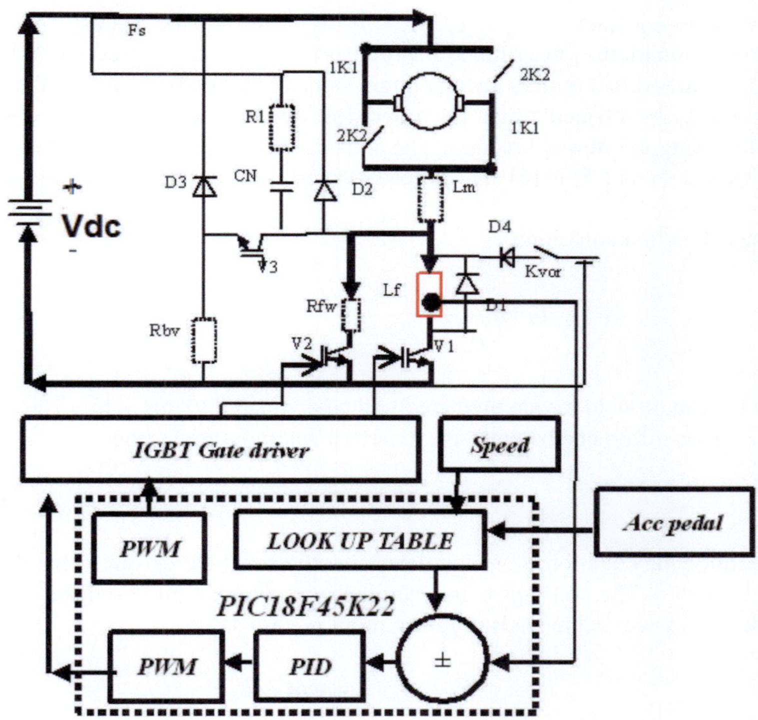

Fig. 11.1 FQDC in fields weakening mode

11.2 Methodology

The sweet spot between the field current that produces the least back EMF and the field current that generates the highest torque can be discovered using GD. The required general control strategy for parallel mode consists of a block diagram and flow chart that are conceptually identical to those in [5]. The torque and speed of the motor serve as important inputs to the GD controller. The controller's output is dependent on these two numbers. The input for the GD operator is the value of the DC motor's speed. An equation that takes speed into consideration suggests the optimal area for field current and torque.

The General equation is:

$$\text{Range}_{I\text{field}} = \frac{V_{\text{dc}}}{\text{Speed}} \times C \tag{11.1}$$

where C is the constant.

The equation for the fitness function provides a more refined expression. When the GD algorithm is run, it returns the maximum value of the field current that fits within the previously established range. Each iteration's success will be measured against this value using the fitness function. The equation of fitness function is determined from Eqs. (11.1)–(11.5) in [6] and in steady-state case is described by the equations below:

Fitness function equation,

$$T_d = \frac{V_{\text{batt}} - K_v I_f \omega}{R_a + \frac{(R_f w_a x R_f)}{R_{fw} + R_f} K_{fw}} \times K_t I_f \tag{11.2}$$

The maximum field torque predicted will be compared to this value. The error is produced by a difference between the expected and measured torque.

$$E_{rr} = E_{\text{val}} - \text{Act}_{\text{val}} \tag{11.3}$$

The following equation is used to determine the new field current value that will be implemented. The constant is the learning rate; a higher number leads to larger steps, while a lower value leads to fewer, more precise steps.

$$F_{\text{new}} = F_{\text{prev}} - (\text{Constant} \times E_{rr}) \tag{11.4}$$

Once the new value of field current has been calculated, the new value is assigned as previous value for the next calculation.

$$F_{\text{prev}} = F_{\text{new}} \tag{11.5}$$

The new value will once again be put through the fitness function test. The number of iterations refers to the predetermined number of times the procedure will be repeated. Once the predetermined maximum number of cycles has been reached, the GD algorithm will pause until the next cycle is initiated. The GD controller will be re-engaged by a change in speed. When GD stops, the best torque has been located, and the optimal field current value in relation to the motor speed will have been determined. The expert system algorithm supports the GD method for configuring LUT, and it operates similarly to how it does in parallel mode, as shown in the flowchart in [7].

11.2.1 Simulation Model and Simulation Result

The simulation model offers fast, economical, time-saving, and safe methods for testing any system. First, a simulation model, as in Fig. 11.2, that contains gradient descent algorithms, is incorporated with expert system algorithms, and is associated with all the mathematical equations for the establishment of a look-up table for the field weakening operation of the FQDC.

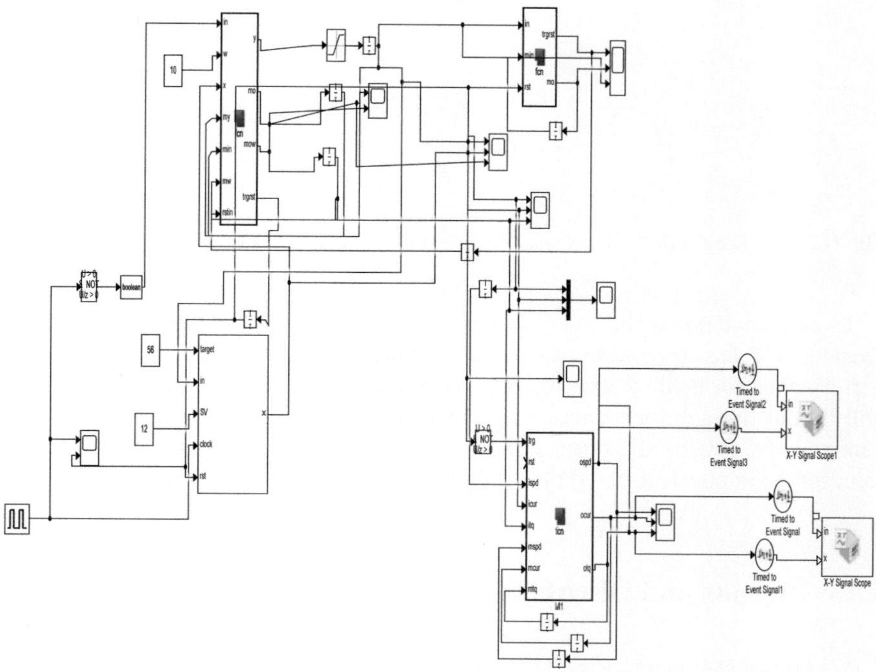

Fig. 11.2 Field weakening LUT control algorithms using GD and expert system

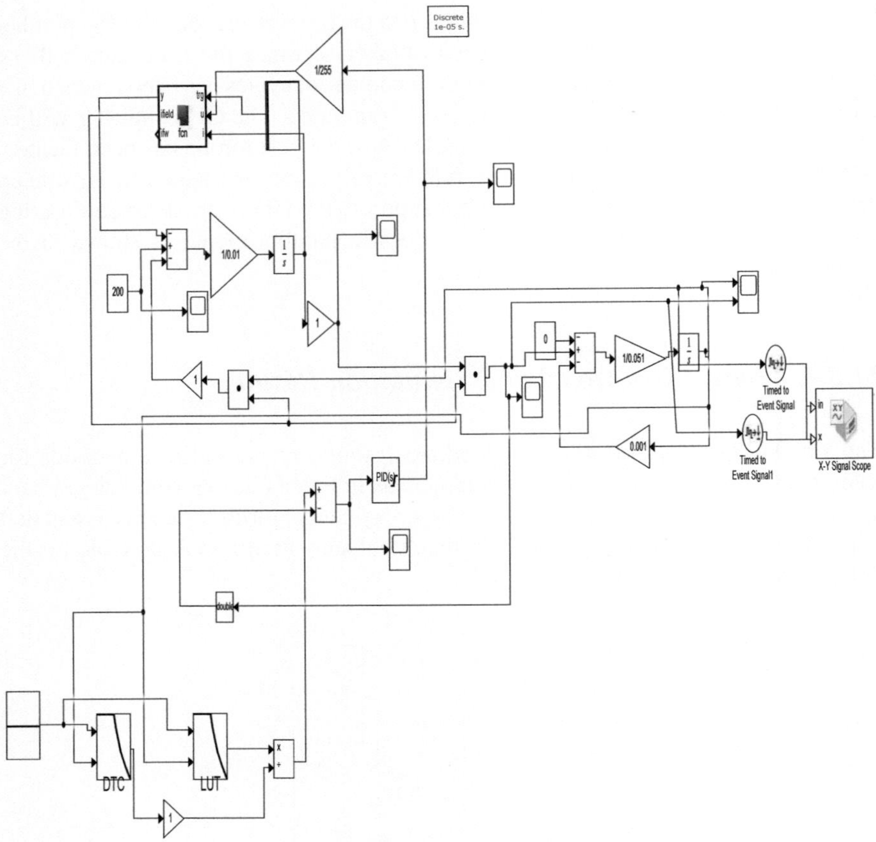

Fig. 11.3 The simulation model of small kilowatt motor using mathematical model

Upon completion of the establishment of the LUT, as shown in Fig. 11.3, the newly created LUT has to be tested to check its functionality, correctness, and expected performance for field weakening mode operation of the FQDC. Simulation models will be put through their paces throughout the testing phase. These models were generated by solving the linear differential equation of a DC series motor and are represented in physical-based modeling using a MATLAB/Simulink model.

11.3 Results and Discussion

The results of setting up a look-up table are shown in Figs. 11.4 and 11.5. Figure 11.5 shows the torque, speed, and field current while under test using gradient descent and expert system algorithms. While the motor speed is set constant and the field current

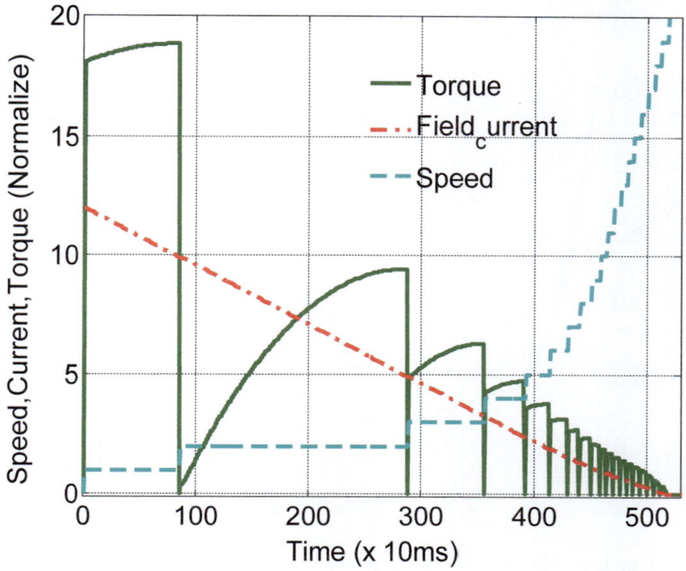

Fig. 11.4 Field weakening speed, torque and field current

is varied to obtain the maximum torque. Once the maximum torque is achieved, a new motor speed is set to find the best field current to obtain the highest motor torque, and the process continues. In Fig. 11.4, for example, for a speed of 1 rad/s, the best field current to achieve the maximum torque is 10 A, while for a speed of 2 rad/s, the best field current to achieve the maximum torque is 5 A.

Figures 11.5 and 11.6 are the results in the form of LUT for the field current and torque versus motor speed, which were extracted from the previous work into a new figure.

The result is shown in Fig. 11.7. On the first test, three simulations using a 0.65 kW motor were done to test the load effect during field weakening using LUT, without LUT, and when no field weakening takes place. When utilized in conjunction with field weakening, the LUT allows for maximum motor speed (in green). The highest speed is achieved when the vehicle is run in field weakening mode with LUT as part of the control system. When field weakening is performed without the use of a LUT, the best possible field current is not attained, resulting in reduced speed (blue). The middle speed is achieved when the system experiences field weakening mode but without using look up table. The lowest speed is when no field weakening is used and is run only in driving mode.

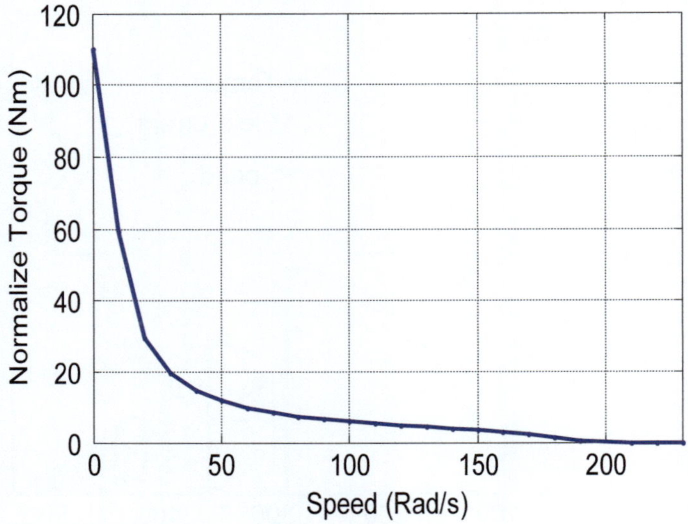

Fig. 11.5 Look up table of torque versus speed for field weakening

Fig. 11.6 Look up table of field current versus speed during field weakening

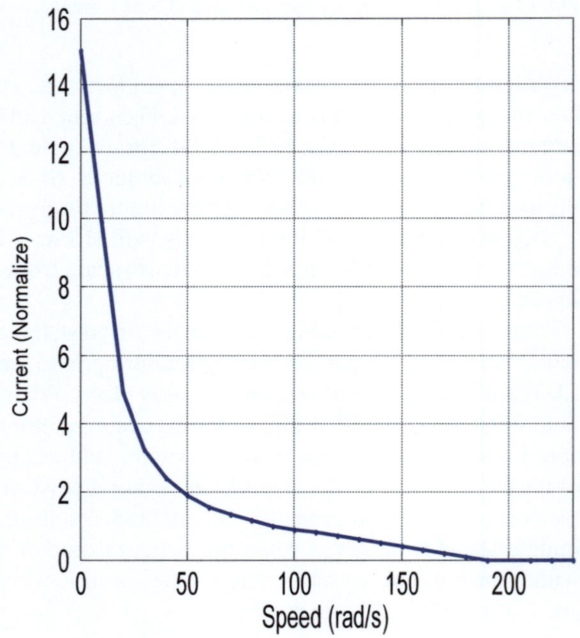

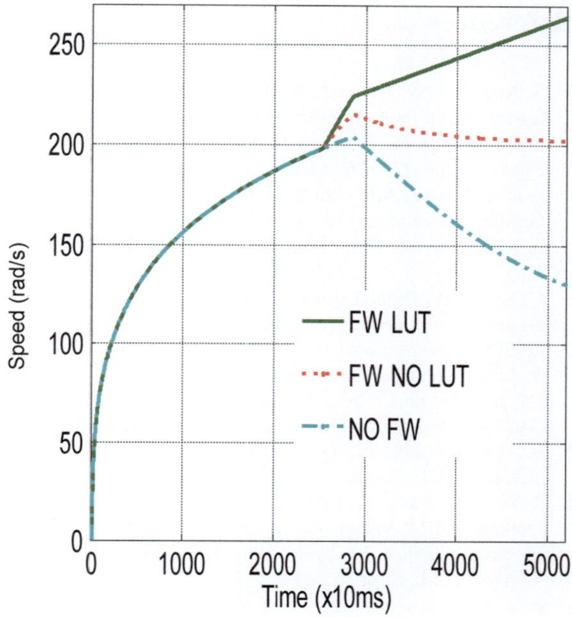

Fig. 11.7 Motor speed in field weakening using LUT, without LUT and no field weakening

11.4 Conclusion

The GD control algorithm and DC motor model have been designed and developed using the MATLAB Simulink software. Prior to the testing in field weakening mode, a LUT was developed, tested, and evaluated using GD. The simulation results of the motor speed have been thoroughly explained and elucidated. The GA LUT performance was evaluated in terms of speed when loaded and the field weakening mode is activated. The simulation results show that GD LUT could change the field current reference to a suitable value according to the speed of the motor. Based on the encouraging findings obtained from the GD LUT simulation results, it is proven that the GD control algorithm and the LUT are robust. For future work, particle swarm optimization (PSO) could be tested to setup the LUT. The integration of an expert system algorithm and a fitness function in a complete simulation model leads to the construction of a look-up table of speed versus field current and speed versus torque, which is then used to evaluate the performance. Using LUT, the FQDC may lessen the back-emf, allowing the vehicle to go faster. To sum up, enhanced LUT performance in field weakening mode is achieved by increasing torque output. In its intended role as an optimization tool, the GD was able to correctly determine the optimal field current for maximum torque, allowing the development of a look-up table for the respective mode. As can be seen from the above comments, DC-driven EVs with series motors have a great deal of promise for use in EVs because of their simple design, cheap cost, and outstanding controllability.

References

1. S. Arof, N.H.N. Diyanah, N.M. Noor, M. Rosyidi, P.A. Mawby, H. Arof, Genetics Algorithm for setting up look up table for parallel mode of new series motor four quadrants DC chopper, in *Progress in Engineering Technology: Automotive, Energy Generation, Quality Control and Efficiency*, pp. 155–167 (2019). https://doi.org/10.1007/978-3-030-28505-0_12
2. A.R.A. Saharul Arof, Shauqee, M. Rosyidi, N.H.N. Diyanah, P. Mawby, H. Arof, E. Noorsal, Gradient descend for setting up a look-up table of series motor four quadrants drive DC chopper in parallel mode. Prog. Eng. Technol. III (2021). https://doi.org/10.1007/978-3-030-67750-3_16
3. C. Lai et al., PMSM drive system efficiency optimization using a modified gradient descent algorithm with discretized search space. IEEE Trans. Transp. Electrification **6**(3), 1104–1114 (2020). https://doi.org/10.1109/TTE.2020.3004463
4. S. Arof, N.H.N. Diyanah, P.A. Mawby, H. Arof, E. Noorsal, Series motor four quadrants drive DC chopper part3: field weakening mode. IOP Conf. Ser. Mater. Sci. Eng. **1127**(1), 012028 (2021). https://doi.org/10.1088/1757-899X/1127/1/012028
5. S. Arof, J.A. Jalil, N.M. Yaakop, P.A. Mawby, H. Arof, Series motor four quadrants drive DC chopper part1:overall. Int. Conf. Power Electron.
6. S. Wang, J. Kang, M. Degano, A. Galassini, C. Gerada, An accurate wide-speed range control method of IPMSM considering resistive voltage drop and magnetic saturation. IEEE Trans. Ind. Electron. **67**(4), 2630–2641 (2020). https://doi.org/10.1109/TIE.2019.2912766
7. S. Arof, N.M. Noor, P. Mawby, H. Arof, E. Noorsal, Embedded system using a PIC microcontroller for series motor four quadrants drive DC chopper controllers for the application in electrical vehicles, in *Progress in Engineering Technology IV. Advanced Structured Materials*, ed. by M.H. Abu Bakar, M.A.H. Abdul Razak, A. Öchsner, vol. 169 (Springer, Cham, 2022). https://doi.org/10.1007/978-3-030-93250-3_3

Chapter 12
Low Harmonics Plug-In Home Charging Electric Vehicle Battery Charger Utilizing Multi-level Rectifier, Zero Crossing and Buck Chopper: Application of Zooming Circuit to Improve the Close Loop Control of Battery Voltage Charging

Norramlee Mohamad Noor, Halina Hassan, Saharul Arof, Norhaslina Ibrahim, Emilia Noorsal, Hamzah Arof, and Philip Mawby

Abstract The low harmonics home charging electric vehicle battery charger for DC drive electric car that makes use of multi-level rectifier, zero crossing, and buck chopper is discussed in this paper. The emphasis is on zooming circuits which consist of voltage divider, offset, subtraction, amplifier circuits and embedded controller. It

N. M. Noor (✉) · H. Hassan · S. Arof · N. Ibrahim
Electrical, Electronic and Automation Section, Universiti Kuala Lumpur, Malaysian Spanish Institute Kulim Hi-Tech Park, Kulim, Kedah, Malaysia
e-mail: noramlee@unikl.edu.my

H. Hassan
e-mail: halina@unikl.edu.my

S. Arof
e-mail: saharul@unikl.edu.my

N. Ibrahim
e-mail: norhaslina@unikl.edu.my

S. Arof · E. Noorsal
Center for Electrical Engineering Studies, Universiti Teknologi MARA, Permatang Pauh Campus, Permatang Pauh, Pulau Pinang, Malaysia
e-mail: emilia.noorsal@uitm.edu.my

H. Arof
Engineering Department, Universiti Malaya, Jalan Universiti, Kuala Lumpur, Malaysia
e-mail: ahamzah@um.edu.my

S. Arof · P. Mawby
School of Engineering, University of Warwick, Coventry, UK
e-mail: P.A.Mayby@warwick.ac.uk

T. A. A. Razak et al. (eds.), *IT Application for Sustainable Living 2*,
SpringerBriefs in Applied Sciences and Technology,
https://doi.org/10.1007/978-3-031-93198-7_12

is used to improve the battery voltage charging performance in close loop feedback control. MATLAB/Simulink software simulates and tests the circuit and its control algorithm. The results suggest that the zooming circuit improves the battery charging voltage in closed loop control.

Keywords Zooming circuit · EV · Battery charger · FQDC

12.1 Introduction

Hydrocarbons pollute the environment and cause global warming, melting icebergs, and raising sea levels. Electric vehicles (EVs) can reduce global hydrocarbon emissions. An EV needs to recharge to restore its battery power. Charging can happen while the EV is moving or when it is at a standstill. The EV can be charged at a charging station or at home when immobile. For charging at home, a low-harmonic type of battery charger is needed. There are several topologies of battery chargers that consist of the multilevel and non-multi-level types available in the literature [1]. The dynamic process during the charging represented by algorithm associated with a circuit model [2]. There is also a study on how EV uncertainty affects the network's imbalance in a distribution grid with greater penetration [3]. An efficient energy transfer system for charging lithium-ion battery-powered electric vehicles has previously been suggested. The controller circuit compares the reference and measured output values to input errors to determine the necessary duty ratio value. The proposed charging technique was successfully tested for the V2V energy transfer system using the MATLAB/Simulink simulation software [4]. At the end or at the output of this battery, the charger is connected to batteries to control the charging voltage and current while charging. Batteries voltage show significant voltage differences at 20% of the total state of charge (SOC), while the remaining 80% show no significant voltage changes. This causes difficulties in controlling the voltage charging. There are several methods of voltage and current control techniques available in the literature. Zooming technique that consists of combination of circuits and algorithm is needed as mentioned in [5, 6]. The fuzzy logic used as a controller for constant voltage and constant current resulted in a lower error percentage [7]. A newer topology using an inductor double capacitor (LCC) is used for constant voltage and constant current, which improves efficiency. The maximum reaches 92.1% [8]. SOC, voltages, and currents of batteries are used as part of control loops to determine the state of the switches, and the duty cycle of the switches based on PI regulators. The proposed control strategy resulted in more balanced paths and completed the equalization process [9]. The phase-shifting control implemented for constant current and constant voltage allows working parameters to be independent of the mutual inductance of the coupling, making the system highly adaptable. Simulation results verify the rationality of the proposed topology [10].

The development of an advanced DC brushed motor with 55 kW power output, 92% efficiency, and 13 V operating voltage boosted the interest in DC drive EC research. A new design of a four-quadrant DC chopper (FQDC) for series motors, as shown in Fig. 12.1, offers more chopper operation to replace the conventional H-bridge. Following after, an EC battery charger, which includes a charging control algorithm, a state of charge estimator, and zero-cross hardware, was studied. This paper continues the discussion on the zero cross, multilevel rectifier, and buck chopper plug-in home charging battery charger as shown in Fig. 12.1. The emphasis is on the zooming circuit specifically applied to the proposed battery charger that controls the voltage and current. This zooming circuit is controlled by the battery management system (BMS), which makes battery chargers safe, reliable, and cost-effective while maintaining battery safety and life. During the generator, regenerative [11], and resistive braking [12] modes, the FQDC charges the batteries while the EV moves. The battery charger circuit charges the batteries when the vehicle stops.

The complete battery charging circuit, charging process, its control algorithms, and the effect of the uncontrolled voltage that can cause high instant charging current to have been discussed in [11]. Figures 12.2 and 12.3 show examples of an extension of the macro charging circuits that can be connected to the buck chopper of a proposed battery charger that allows series and parallel battery charging.

Figure 12.4 shows the pattern of state of charge and open circuit voltage characteristics while charging until it is fully charged. Careful observation of the battery SOC versus open circuit voltage shows that some batteries show significant voltage differences at 20% of the total SOC, while the remaining 80% show no significant voltage changes.

If a microcontroller is built into an A/D converter with low resolution, it might not detect these small voltage changes. The controller that receives this incorrect battery voltage reading may either ignore or misinterpret it. This results in poor charging performance that deters the full charging of the batteries. This paper discusses the zooming circuit and how it enlarges this small but crucial signal to improve the battery charging performance.

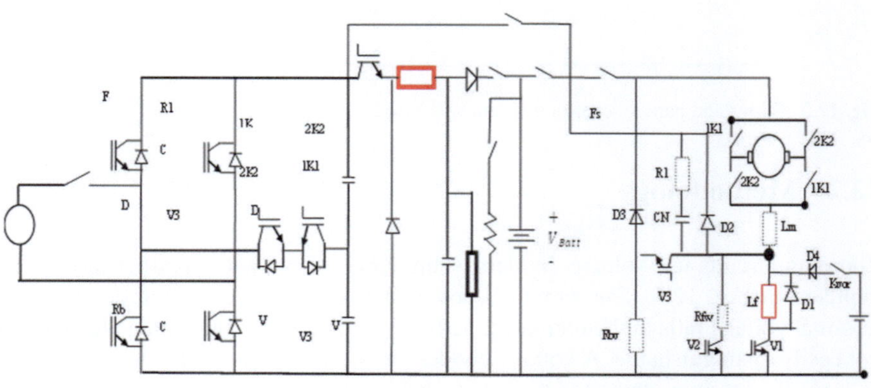

Fig. 12.1 Integrated FQDC chopper and battery charger

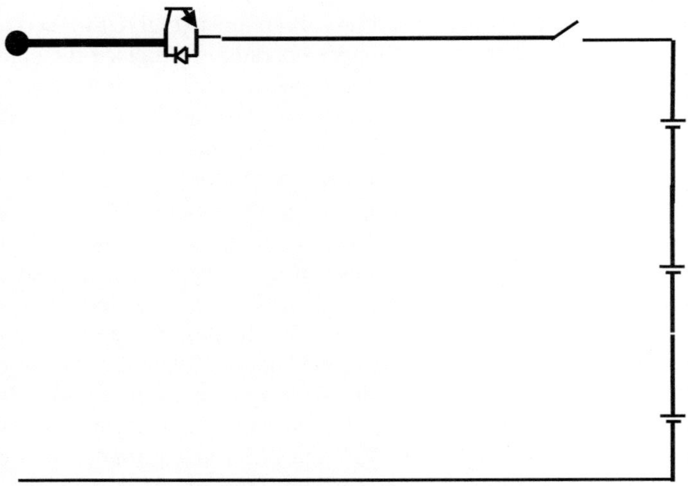

Fig. 12.2 Series charging circuit

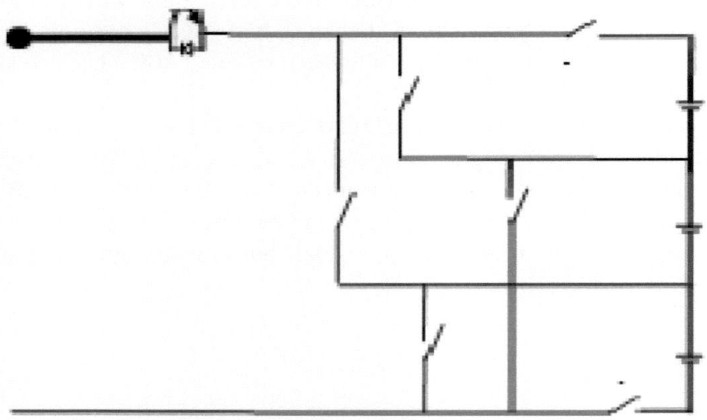

Fig. 12.3 Series and parallel charging circuit with single supply

12.2 Methodology

Zooming circuits use voltage dividers, subtractors, offset, and amplifier circuits as depicted in Fig. 12.6. This circuit becomes active when the battery open circuit terminal voltage falls in the region where SOC increases, and the voltage increase is not easily distinguishable. A voltage divider circuit is used to divide the voltage that exceeds the op-amp integrated circuits or A/D converter. The subtractor circuit has two inputs, one of which is the reference voltage, set at the point where the voltage shows no obvious increase. The other input is the open-circuit battery voltage input.

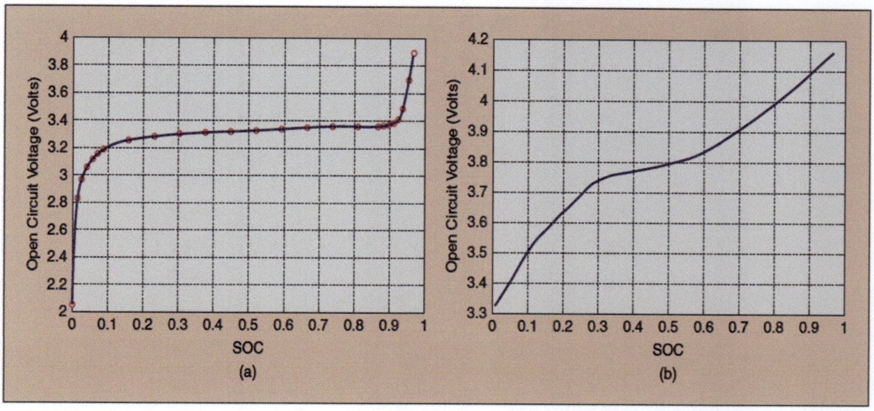

Fig. 12.4 Battery voltage pattern at fully charge

This circuit feeds the offset circuit, or amplifier voltage circuit. The voltage offset circuit shifts the voltage to the designer's chosen range. The amplifier amplifies the voltage to match the A/D converter's resolution on a microcontroller or embedded system. The controller has a control algorithm running close loop feedback control that can finely control the voltage and current in a closed loop system. The input reference for the close loop feedback control is a running algorithm regulated accordingly to match the zooming circuit feedback signal.

In embedded systems such as PIC microcontroller in Figs. 12.5 and 12.6, algorithms are coded, stored, and executed.

This allows close loop controller tuning to the imitate system. Such an example of numerical computation using Taylor series for fault diagnose and online system tuning and optimization for driving mode operation of FQDC is reported in.

12.2.1 *Simulation Model of Battery Charger, Charging Controller, and Zooming Circuit*

Figure 12.7 shows a MATLAB/Simulink simulation model of zero crossing, MLR, buck chopper, and charging controller for testing the battery charger and control algorithm. Figure 12.8 shows the associated voltage divider, subtract, offset, and amplifier simulation circuits. The voltage terminal voltage can be obtained from the internal battery simulation model or based on the state of charge estimator [2] or charging current according to Eq. 12.1.

$$I_c = \frac{v_c - V_T}{R\ \mathrm{batt}} \tag{12.1}$$

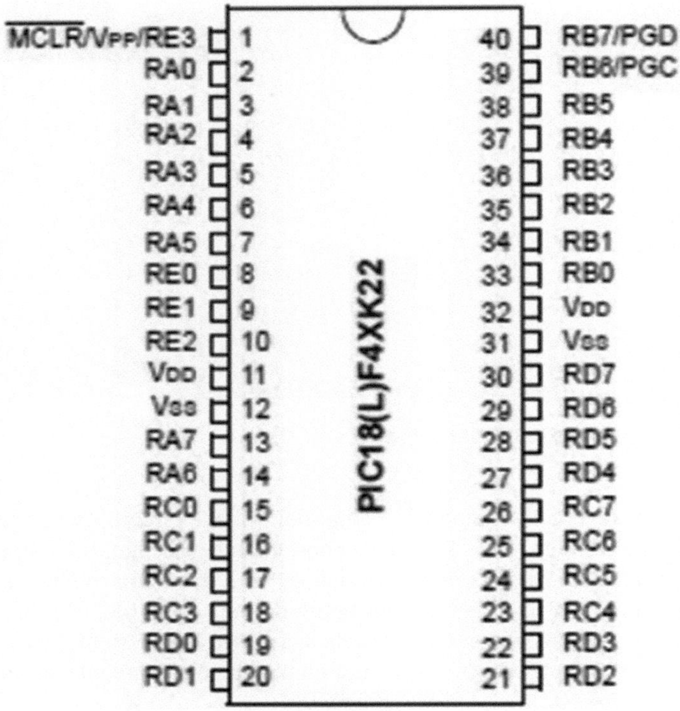

Fig. 12.5 PIC microcontroller

12.3 Results and Discussion

Figure 12.9 shows that the voltage divider circuit divides the terminal voltage into about half of its volt during charging.

Figure 12.10 shows the voltage divider output voltage (red), being offset using an offset circuit, and the result for the subtractor/offset circuit (– 5.5 V) now starts at 0. Figure 12.11 shows the amplifier circuit that amplifies the offset input. The output is the result of multiplication by the amplifier.

Figure 12.12 shows the original voltage and the final voltage of the zooming circuit. Figure 12.13 shows the zooming circuit control performance for controlling the battery charging voltage at 16.2 V, which falls under the region where the voltage increase is hardly distinguishable. The zooming effect circuit performs better in close loop control of battery voltage control, as it shows less overshoot and zero steady state error. The harmonics and THD results have been discussed in [10] and are not covered in this paper.

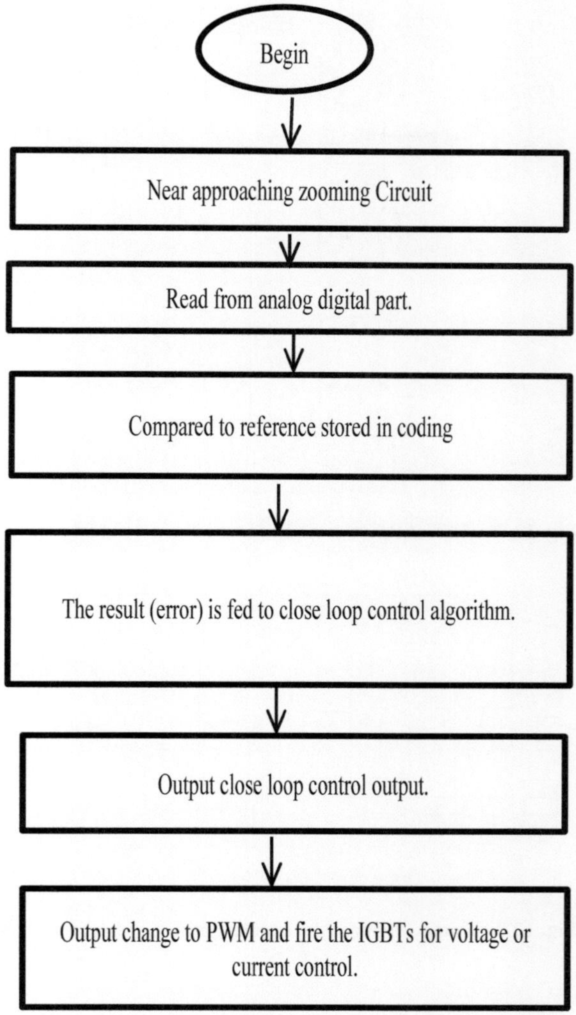

Fig. 12.6 Block diagram of the operation

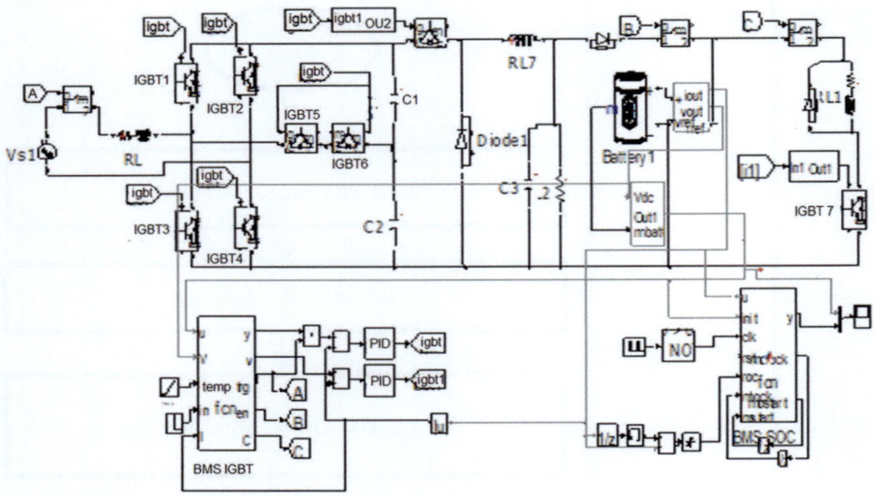

Fig. 12.7 MATLAB/Simulink BMS and charger model

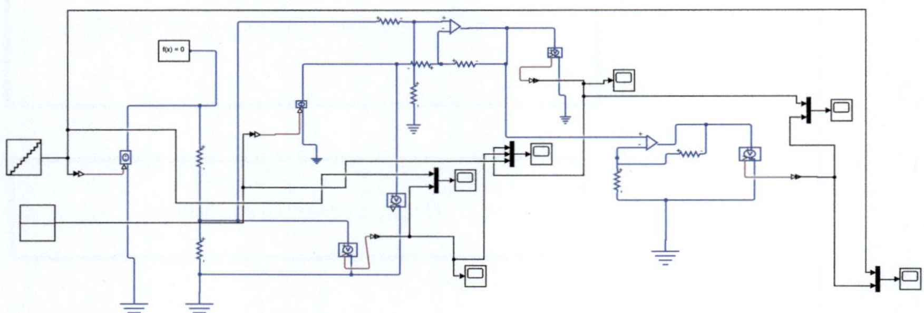

Fig. 12.8 Voltage divider, offset and amplifier circuit for zooming circuit

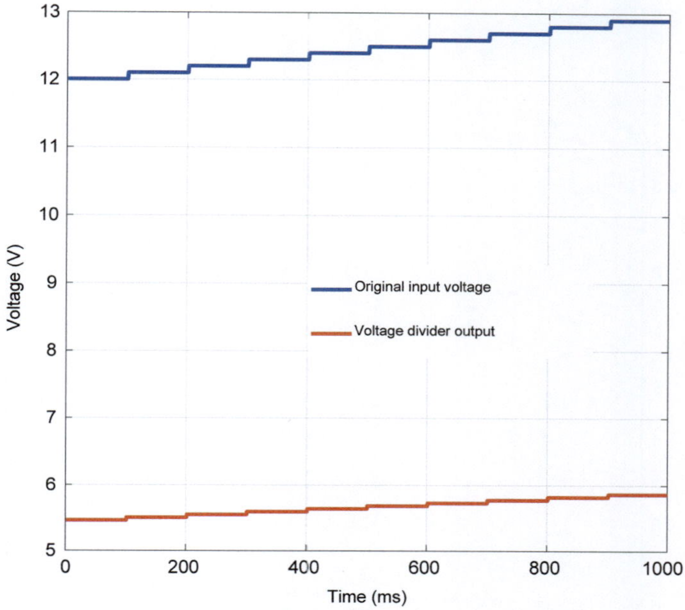

Fig. 12.9 Voltage divider input and output voltage

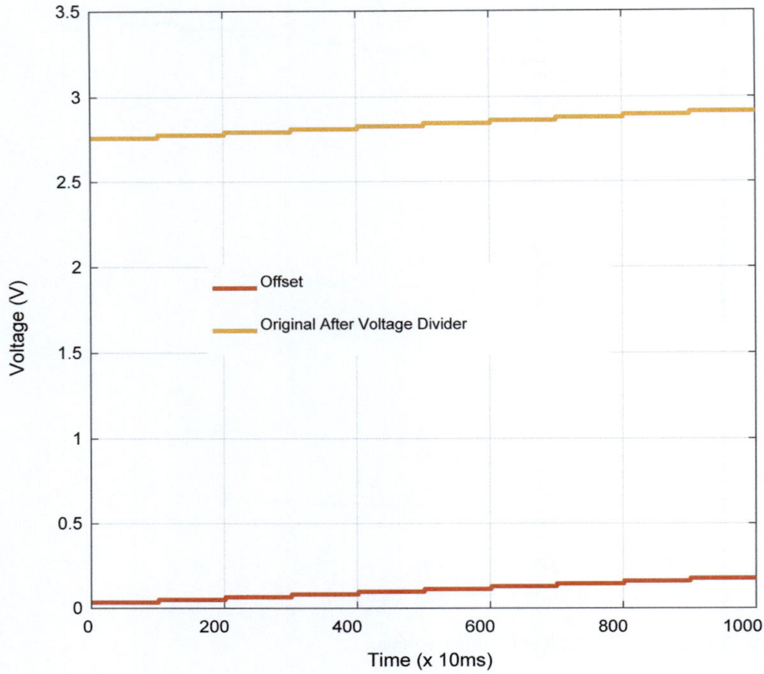

Fig. 12.10 Input and output of offset/subtract circuit

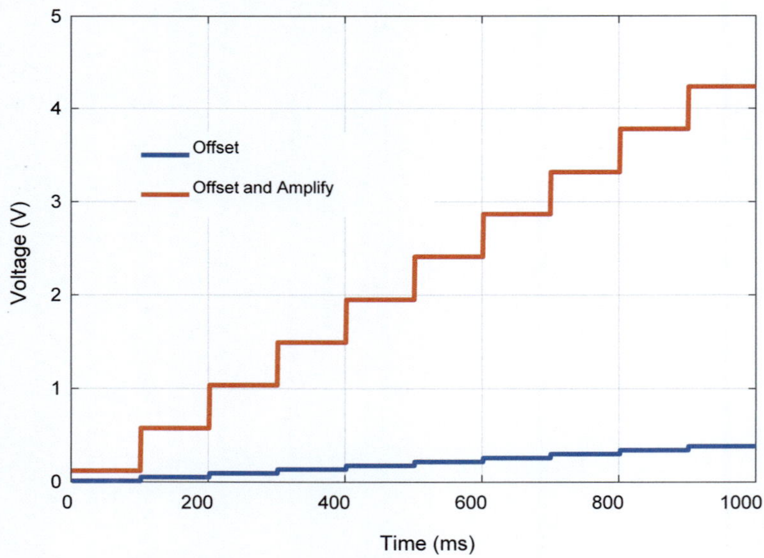

Fig. 12.11 Input signal of offset/subtract circuit and output signal after been amplified

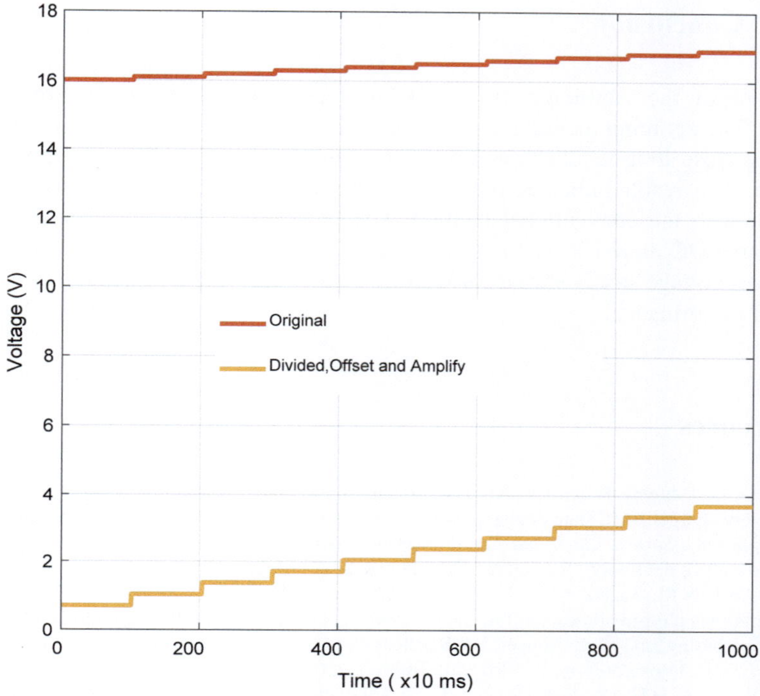

Fig. 12.12 The input and output of zooming circuit

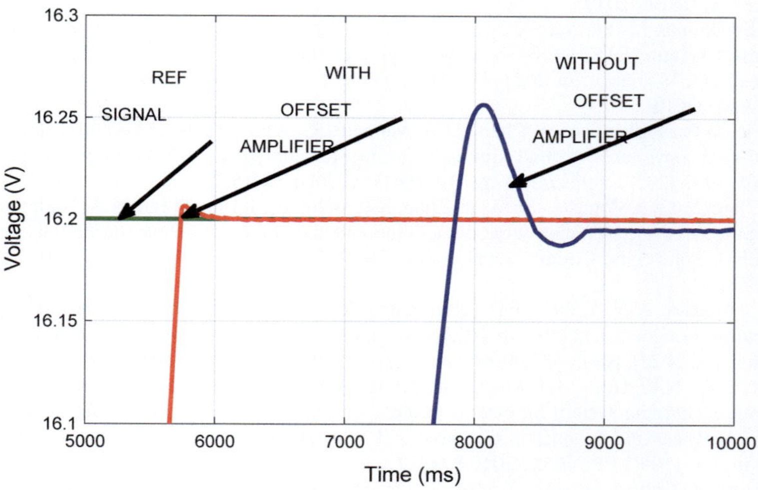

Fig. 12.13 Comparison of using zooming circuit and not using zooming circuit

12.4 Conclusion

In this paper, the zooming circuit is used to improve the close loop control performance. The zooming includes analog circuits associated with embedded controllers running close loop feedback control algorithm to process and control the overall system. The result indicates improved performance in close loop control when compared to the conventional method. The proposed battery charging controller can charge DC-drive EV batteries on static AC power, while the zooming technique improves voltage steady state error while providing fast response and low overshoot control performance.

References

1. S. Arof, P. Mawby, H. Arof, E. Noorsal, Low harmonics plug-in home charging electric vehicle battery charger BMS state of charge balancing using priority and sequencing technique, in *IOP Conference Series: Earth and Environmental Science*, vol. 1261. International Postgraduate Conference for Energy Research 2022 (IPCER-2022) (2022)
2. S. Arof, N.M. Sazali, M.N. Noor, P.M. Amirah, E. Noorsal, A.A. Bakar, Y.M. Ali, Low harmonics plug-in home charging electric vehicle battery charger utilizing multi-level rectifier, zero crossing, and buck chopper: BMS battery charging control algorithm. Prog. Eng. Technol. III (2021). https://doi.org/10.1007/978-3-030-67750-3_18
3. M.R. Islam, H.H. Lu, M.J. Hossain, L. Li, Compensating neutral current, voltage unbalance and improving voltage of an unbalanced distribution grid connected with EV and renewable energy sources, in *2019 22nd International Conference on Electrical Machines and Systems (ICEMS)* (IEEE, 2019)
4. N. Sivakumar, M. Saravanan, S.C. Raja, S.A.E. Xavier, Analysis the performance of energy transfer system (ETS) for EV battery charging method using non-isolated SEPIC converter-based CCCV control, in *2023 9th International Conference on Electrical Energy Systems (ICEES)* (IEEE, 2023)
5. M. Reza Pazhouhandeh, M. Chang, T.A. Valiante, R. Genov, Track-and-zoom neural analog-to-digital converter with blind stimulation artifact rejection. IEEE J. Solid-State Circuits **55**(7), 1984–1997 (2020). https://doi.org/10.1109/JSSC.2020.2991526
6. X. Tang, S. Li, X. Yang, L. Shen, W. Zhao, R.P. Williams, J. Liu, Z. Tan, N.A. Hall, D.Z. Pan, N. Sun, An energy-efficient time-domain incremental zoom capacitance-to-digital converter. IEEE J. Solid-State Circuits **55**(11), 3064–3075 (2020). https://doi.org/10.1109/JSSC.2020.3005812
7. F.C. Arsandi, M.Z. Efendi, F.D. Murdianto, Constant current constant voltage for precise lithium-ion battery charging, in *2022 International Electronics Symposium (IES)* (Surabaya, Indonesia, 2022), pp. 48–53. https://doi.org/10.1109/IES55876.2022.9888703
8. N.T. Diep, N.K. Trung, T.T. Minh, Control the constant current/voltage charging mode in the wireless charging system for electric vehicle with LCC compensation circuit, in *2019 IEEE Vehicle Power and Propulsion Conference (VPPC)* (Hanoi, Vietnam, 2019), pp. 1–5. https://doi.org/10.1109/VPPC46532.2019.8952272
9. X. Liu, Z. Gao, X. Huang, Y. Zou, Large equalization current control strategy for series connected battery packs based on buck-boost converter, in *2018 International Power Electronics Conference (IPEC-Niigata 2018-ECCE Asia)* (Niigata, Japan, 2018), pp. 3455–3459. https://doi.org/10.23919/IPEC.2018.8507560
10. D. Pavković, M. Hrgetić, A. Komljenović, V. Smetko, Battery current and voltage control system design with charging application, in *2014 IEEE Conference on Control Applications*

(CCA) (Juan Les Antibes, France, 2014), pp. 1133–1138. https://doi.org/10.1109/CCA.2014.6981481

11. S. Arof, N.H.N. Diyanah, P. Mawby, H. Arof, N. Mohd Yaakop, Low harmonics plug-in home charging electric vehicle battery charger utilizing multi-level rectifier, zero crossing and buck chopper: part 1: general overview, in Progress in Engineering Technology,pp. 103–108. https://doi.org/10.1007/978-3-030-28505-0_9
12. S. Arof, M.R. Ahmad, N.M. Yaakop, P. Mawby,H. Arof, E. Noorsal, A.A. Bakar, Y. Mohd Ali, Low harmonics plug-in home charging electric vehicle battery charger utilizing multilevel rectifier, zero crossing, and buck chopper: state of charge estimator using current integration algorithm for embedded system (2021)

Chapter 13
Development of Genetics Algorithm Look-Up Table in the Field Weakening Mode of a Series Motor Four Quadrants Drive DC Chopper

Nurazlin Mohd Yaakop, Saharul Arof, Azmi Hassan, Norzalina Othman, Philip A. Mawby, Hamzah Arof, and Emilia Noorsal

Abstract This paper presents the implementation of an AI strategy genetic algorithm (GA) in the field weakening DC motor controller of a four quadrant drive DC chopper (FQDC). The field weakening mode of the FQDC for series motor could fail which causes the torque and motor speed decrease if the FQDC cause an incorrect field motor current to the motor. In the presence of the worst maximum magnetic field due to increase motor speed demand above base speed, a field current imbalance condition occurs in the motor control system. Hence, a genetic algorithm is used to find the most optimized firing angle for different equal or unequal armature current and field

N. M. Yaakop (✉) · S. Arof · A. Hassan · N. Othman
Electrical, Electronic and Automation Section, Universiti Kuala Lumpur, Malaysian Spanish Institute Kulim Hi-Tech Park, Kulim, 09000, Kedah, Malaysia
e-mail: nurazlin@unikl.edu.my

S. Arof
e-mail: saharul@unikl.edu.my

A. Hassan
e-mail: azmi.hassan@unikl.edu.my

N. Othman
e-mail: norzalina@unikl.edu.my

P. A. Mawby
University of Warwick, School of Engineering, CV47AL, Coventry, UK
e-mail: P.A.Mawby@warwick.ac.uk

H. Arof
Engineering Department, Universiti Malaya, Jalan Universiti, 50603, Kuala Lumpur, Malaysia
e-mail: ahamzah@um.edu.my

E. Noorsal
Electrical Engineering Studies, College of Engineering, Universiti Teknologi MARA Cawangan Pulau Pinang, Permatang Pauh Campus, Permatang Pauh, 13500, Pulau Pinang, Malaysia
e-mail: emilia.noorsal@uitm.edu.my

T. A. A. Razak et al. (eds.), *IT Application for Sustainable Living 2*,
SpringerBriefs in Applied Sciences and Technology,
https://doi.org/10.1007/978-3-031-93198-7_13

current of the two parallel switches, at which the highest torque will be present in the system. Simulations are performed in the MATLAB/SIMULINK environment.

Keywords DC-DC chopper · Electric vehicle · DC motor · GA · FQDC

13.1 Introduction

The resurgence of interest in electric vehicles (EVs) has ignited a fresh exploration of DC motors as a viable power source, especially with the emergence of highly efficient DC motors capable of running at lower voltage levels [1]. This renewed enthusiasm has ushered in innovative devices designed to support this vision, among which the four quadrants DC chopper (FQDC) holds a prominent position, offering multiple operational modes [2]. The performance of EVs equipped with FQDCs and series DC motors in diverse driving scenarios, including acceleration, braking, and even regenerative power generation for battery charging, has become increasingly reliant on advanced artificial intelligence (AI) techniques such as fuzzy logic, self-tuning fuzzy systems, neural networks, and adaptive neuro fuzzy inference systems (ANFIS) [3–5]. These AI-driven systems have demonstrated their ability to optimize motor control and enhance vehicle performance across various terrains. In the pursuit of further refining FQDC performance for DC-driven EVs, researchers have turned to the power of artificial intelligence, specifically leveraging techniques like genetic algorithms (GA) [6, 7] and gradient descent (GD) [8]. These methodologies have been employed to optimize specific chopper operational modes, notably the parallel mode, through the creation of a look-up table that correlates the field current with motor speed. This paper extends this optimization journey to encompass another critical chopper operation mode, namely the field weakening mode [9].

13.2 Literature Review

13.2.1 Genetic Algorithm

Genetic algorithms (GAs) [6–12] represent a powerful optimization technique that has found wide-spread application in solving intricate optimization problems and control processes. Their appeal lies in their ability to search for global solutions through mechanisms like mutation, selection, and cross-over. In the realm of electric vehicles driven by DC motors, GAs holds the potential to optimize various operational modes by fine-tuning critical variables such as current, voltage, and torque.

Figure 13.1 provides a visual representation of the current distribution through the primary IGBT V1 and IGBT V2 during the field weakening mode, a topic that has been comprehensively explored in previous research [8]. The field weakening mode is essential for boosting the motor torque to overcome loads or increase motor speed, and this operational mode has been extensively examined elsewhere [8]. The passage of current through the field weakening resistor results in a reduction of the field current, consequently leading to a decrease in back electromotive force (EMF). Simultaneously, the field winding is connected in parallel with the field weakening resistor, diminishing the overall resistance in the circuit. This reduction in total resistance results in an increase in the total motor current.

This paper, however, delves into the establishment of a look-up table (LUT) specifically tailored for optimizing the torque by precisely controlling the field current in relation to motor speed during field weakening mode. It is important to highlight that the relationship between torque, which depends on the armature current and field current, and back EMF and motor speed is inherently nonlinear. Consequently, creating an effective LUT through optimization tools such as genetic algorithms [8], augmented with expert system assistance, emerges as the most promising solution for maintaining superior torque performance.

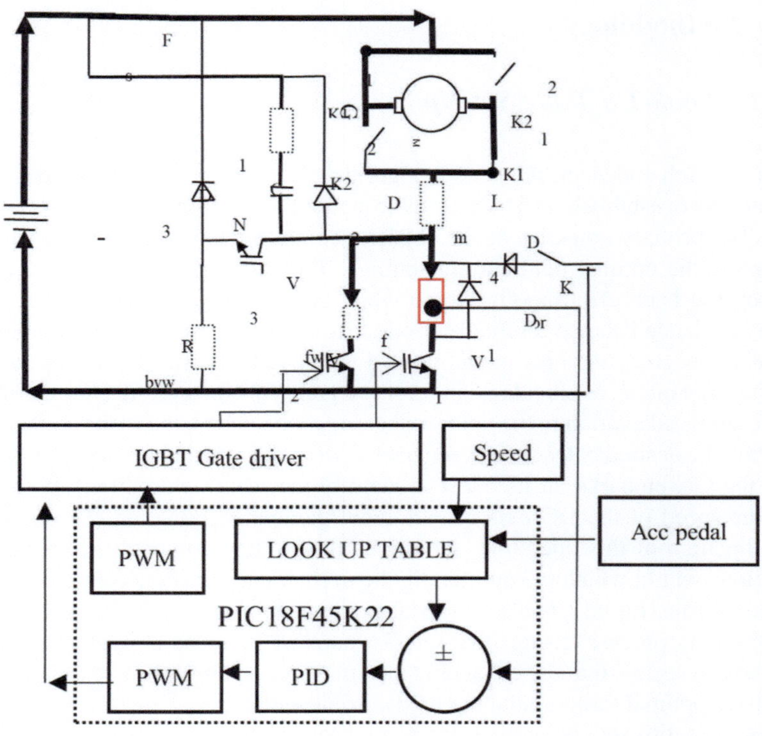

Fig. 13.1 FQDC in field weakening mode

The LUT we propose serves as a crucial reference [6, 8, 13] for the closed-loop control of the field current during field weakening mode. By utilizing genetic algorithms in its creation, we aim to identify the optimal field current settings that yield the highest torque output. This endeavor aligns with the broader objective of enhancing the performance of four quadrants DC choppers in driving electric vehicles, fostering greater efficiency, and contributing to the advancement of sustainable transportation solutions. In the subsequent sections of this paper, we delve deeper into the methodology employed for LUT establishment, simulation results, and the implications for the field weakening mode operation of FQDCs in electric vehicles.

The primary objective of this study is to establish a comprehensive look-up table [6, 8, 13] tailored for the field weakening mode operation of FQDC, with a particular emphasis on utilizing the genetic algorithm for this purpose. By doing so, we aim to unlock the full potential of the FQDC in enhancing the performance of DC-driven electric vehicles, thereby contributing to the sustainable and efficient evolution of the automotive industry.

13.3 Methodology

13.3.1 Look-Up Table Set-Up

In this research endeavor, the genetic algorithm takes center stage, tasked with the intricate task of establishing a look-up table meticulously tailored for field weakening mode. The primary objective of the GA is to comb through a vast parameter space to pinpoint the optimal field current settings. These settings play a pivotal role in reducing the back electromotive force while concurrently maximizing the motor torque, a delicate balance essential for effective field weakening mode operation.

The architecture and operational flow of the GA algorithm mirror those utilized in the parallel mode, as elucidated in prior research [6]. However, the key distinction here lies in its adaptation to the field weakening mode, where the focus shifts toward achieving the desired balance between back EMF reduction and torque optimization.

As the GA embarks on its quest to identify the ideal field current, it crucially takes the speed of the DC motor as its input parameter. Equation (13.1) emerges as the linchpin of this operation, serving as the guiding compass that demarcates the regions within which the optimal field current values are concealed. Beyond its directional role, Eq. (13.1) also undertakes the task of rescaling the search space. This rescaling process ensures that the GA narrows its focus to explore a smaller, more refined region, thereby enhancing the efficiency and precision of the search for the elusive optimal torque-inducing field current settings.

In essence, this section outlines the pivotal role of the GA in the establishment of the LUT for the field weakening mode, shedding light on its mechanism of operation, input parameters, and the mathematical framework that guides its quest for optimal field current values. The subsequent sections of this paper delve into the practical implementation of these concepts and present the results of simulations, showcasing the tangible benefits of this approach in enhancing the performance of FQDCs during field weakening mode operation in electric vehicles. To facilitate the search for the optimal field current, using genetic algorithm, the possible range of field current is determined by motor speed, supply voltage, and a system-specific constant, as shown in Equation (13.1). This range defines the search boundaries for the genetic algorithm to identify the most efficient operating point.

$$[\![\mathrm{Range}]\!]_{I\mathrm{field}} = V_{\mathrm{dc}}/\mathrm{Speed} \times C \tag{13.1}$$

where C is the constant

A refined and more precise equation has been derived from the mathematical expressions in Eqs. (13.1)–(13.4), as described in references [1, 7]. This refined equation, denoted as Eq. (13.2), serves as the fitness function for our optimization process. In our research, we leverage MATLAB to implement this fitness function within a genetic algorithm framework, and the resulting optimized solution is visualized using the MATLAB plot command.

To elucidate the GA process in greater detail, it is important to note that upon activation, the GA algorithm initializes a population of individuals. In our case, this population consists of six genes, each possessing six chromosomes. These six genes are representative of various field current options, and the GA controller, along with its control algorithm, will fine-tune the genetic makeup of these chromosomes to arrive at an optimal solution.

The evaluation of each chromosome's fitness is a critical step in the GA optimization. To assess the strength and suitability of each chromosome, we employ the fitness function defined in Eq. (13.2). This fitness function quantifies how well each individual chromosome aligns with the desired objectives of the optimization problem, thus guiding the GA towards finding an optimal solution.

$$T_d = (V_{\mathrm{batt}} - k_v \times I_f \times \omega)\,/(Rf_w \times R_f)/(R_f w + R_f) \times Ff_w) \times K_t I_f \tag{13.2}$$

The equation in question serves as a critical tool for assessing the performance of individual chromosomes within our genetic algorithm framework. Specifically, this equation calculates the motor torque output, which is of paramount importance in our quest to identify the optimum torque for our application.

In the context of our genetic algorithm, it is imperative to distinguish between the strength of chromosomes, as this factor fundamentally influences the evolutionary process. Chromosomes that exhibit weaker characteristics yield lower values of the output, representing suboptimal solutions. Conversely, strong chromosomes yield higher output values, indicating their effectiveness in achieving the desired torque levels.

Chromosomes that underperform, i.e., those with lower fitness function values, face several potential outcomes within the genetic algorithm. They may undergo destruction, mutation, or have their genetic makeup altered to create space for other, more promising chromosomes with superior fitness functions. The genes responsible for producing high torque in surviving chromosomes engage in a dynamic exchange, leading to the emergence of novel genes and chromosomes. This iterative process eventually culminates in the consolidation of genes from diverse chromosomes into a unified chromosome structure, ensuring a consistent number of genes throughout the algorithm's execution.

The algorithm iterates a predefined number of times, determined either by preset iterations or other criteria. Upon reaching the specified number of iterations, the genetic algorithm halts temporarily, awaiting reactivation for subsequent cycles, which may be triggered by alterations in the ship's speed. The GA controller remains responsive to changes in the operational conditions, thus allowing for dynamic adjustments.

When the genetic algorithm reaches its final iteration and subsequent halt, it identifies the best-performing chromosomes. At this point, the system has successfully pinpointed the optimal field current value corresponding to the motor speed, a critical outcome for our application. It is worth noting that an if–then rules expert system plays a pivotal role in orchestrating the sequence and transitions for evaluating various speed points using the genetic algorithm. This expert system's guidance is instrumental in achieving the most effective selection of field current values for each motor speed, ensuring optimal performance across a range of operating conditions.

13.3.2 Simulation Configuration

This model was constructed to contain a DC chopper, a PID controller, and a GA controller. Developed in the MATLAB/Simulink environment, to create a dynamic representation of a DC series motor. Initially it solved the linear differential equation governing the motor's behavior, and then proceeded to develop a GA.

In this research, a model of a DC series motor with a power rating of 0.6 kW was designed. This model serves as the foundation for the creation of a lookup table using the genetic algorithm implemented in MATLAB/Simulink as shown in Fig. 13.2. This model not only captures the dynamic behavior of the DC series motor but also leverages advanced control strategies such as the PID controller and the GA controller to optimize its performance. This model lays the groundwork for our subsequent investigations, enabling us to explore and refine the motor's operational characteristics and control strategies in a systematic and versatile manner.

In this research, a simulation model of a petite yet robust 0.6 kW motor undergoing rigorous testing is represented, both under load and without. This evaluation takes place within the challenging context of field weakening mode, a critical operational scenario for such motors. The juxtaposition of results with and without the utilization of a look-up table unveils crucial insights into the motor's performance and underscores the impact of advanced control strategies.

Moving on to Fig. 13.2, the initial stages of our GA-based optimization process. This figure showcases the first generation of genes that underwent rigorous testing using our fitness function, meticulously executed through the MATLAB code. In this

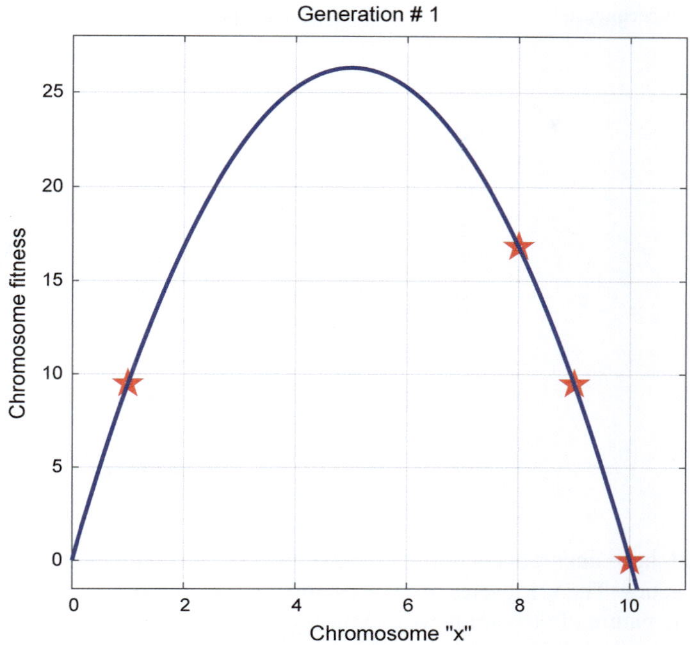

Fig. 13.2 First generation of genetics algorithm

genetic exploration, four out of the six genes emerged prominently, having demonstrated their prowess in the quest for the maximum torque value. Note the remaining two genes are by no means disregarded; instead, they have undergone transformations, merging into the prevailing four genes. The absence of explicit highlighting is primarily due to the intricate interplay and overlapping characteristics within the genetic population, a testament to the dynamic nature of our optimization process.

13.4 Results and Discussion

These results signify the early stages of our investigation, shedding light on the evolutionary dynamics of our genetic algorithm. They set the stage for deeper explorations into gene refinement, chromosome evolution, and the eventual emergence of the most adept genetic configurations. Ultimately, these findings contribute to our overarching goal of optimizing the motor's performance in field weakening mode, a pivotal achievement in the realm of motor control.

Moving forward, transition to Fig. 13.3, representation of our genetic algorithm's evolution. At this stage, just two distinct genes within the genetic population are emerging. It is important to note that the remaining genes, though not explicitly

Fig. 13.3 GA second generation

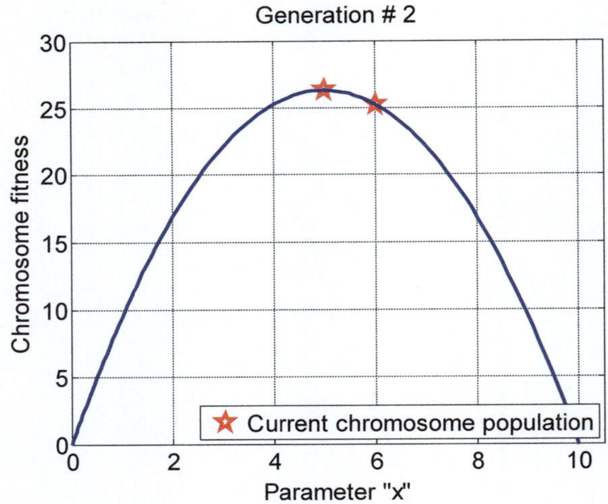

highlighted, have undergone a transformative process, amalgamating into these two dominant genes. The intricacies of this genetic overlap and integration underscore the dynamic nature of our optimization journey.

As we progress to Fig. 13.4, now in the fourth generation of our genetic algorithm exploration. At this advanced stage, evolution has in a remarkable development: the emergence of a single, solitary gene. The absence of additional genes in this representation is primarily a result of extensive overlapping and integration among the genetic constituents.

The convergence towards a reduced number of genes of genetic configurations that exhibit superior fitness in optimizing our target parameter. This streamlined

Fig. 13.4 GA fourth generation

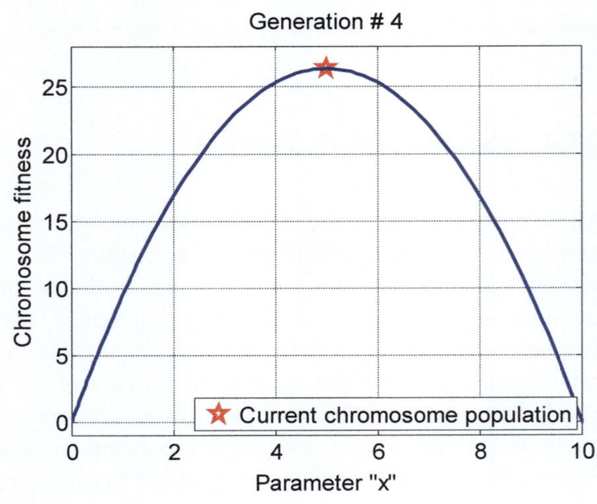

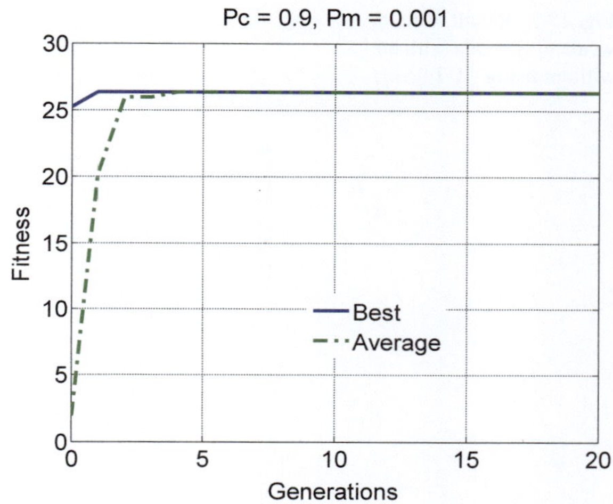

Fig. 13.5 Best and average torque

genetic population signifies the refinement and specialization of genetic makeup, ultimately leading us closer to our research goal of achieving the most adept genetic configuration for our application.

In Fig. 13.5, a comprehensive analysis of genetic algorithm's performance, the fitness function. This figure provides a visualization of both the best and average performance outcomes across various generations of the genetic algorithm.

Figure 13.6 unveils the complete look-up table in the form of a graph, that meticulously maps the motor speed against the corresponding field current values. Within this LUT, we capture the essence of our genetic algorithm's prowess, as it fine-tunes the motor's operation to generate the most favorable torque–speed relationship. This comprehensive table not only showcases the optimal field current settings for a range of motor speeds but also provides a visual representation of the generated torque as a function of speed.

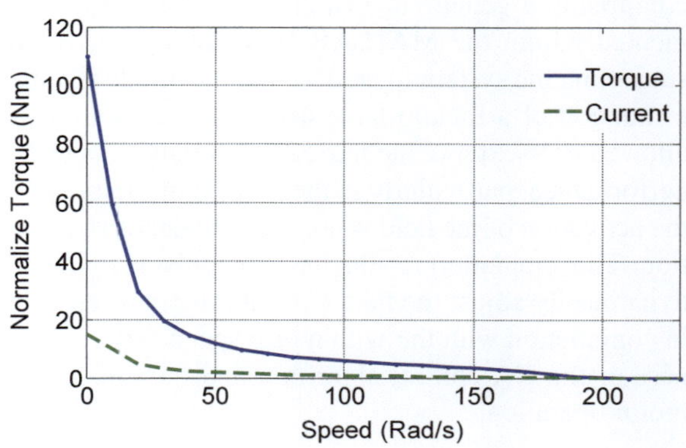

Fig. 13.6 Motor speed and torque versus motor speed look-up table

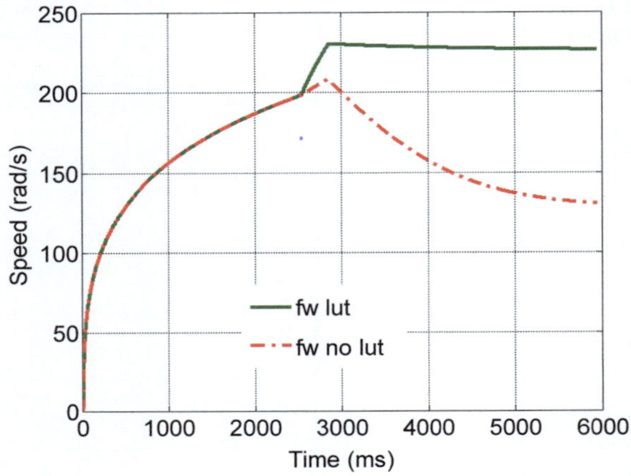

Fig. 13.7 Result of field weakening mode with and without using LUT

Figure 13.7 draws a clear distinction between two scenarios: one with the utilization look-up table optimization and the other without it. The application of LUT optimization leads to significantly improved motor speed performance. This enhancement is particularly pronounced when the motor operates under challenging conditions, such as when the field weakening mode is invoked and load effects come into play. When LUT optimization is not employed, we observe a decline in motor speed when the motor experiences load effects in conjunction with the field weakening mode. This decline is attributed to an incorrect selection of field current at that specific motor speed, which leads to suboptimal performance.

13.5 Conclusion

In conclusion, this research has entailed the design and development of a system comprising a genetic algorithm control algorithm and a DC motor model, implemented within the MATLAB Simulink software environment. Crucially, prior to subjecting the system to field weakening mode testing, the development, testing, and evaluation of a lookup table using the GA optimization technique. This approach allowed us to explore the intricacies of motor control. The analysis of the GA LUT's performance, particularly in the context of motor speed under loaded conditions and the activation of the field weakening mode, has been a central focus of our investigation. The simulation results have unequivocally demonstrated GA LUT's ability to dynamically adjust the field current reference, ensuring optimal motor performance in conjunction with the with motor speed. This adaptability underscores the robustness of the GA control algorithm and the effectiveness of the LUT in optimizing motor operation.

References

1. A.M. Lulhe, T.N. Date, A technology review paper for drives used in electrical vehicle (EV) & hybrid electrical vehicles (HEV), pp. 632–636 (2015)
2. S. Arof, J.A. Jalil, N.M. Yaakop, P.A. Mawby, H. Arof, Series motor four quadrants drive dc chopper part1:overall (2014)
3. S. Arof, A.K.M. Khairulzaman, J.A. Jalil, P.A. Mawby, H. Arof, Artificial intelligence controlling chopper operation of four quadrants drive DC chopper for low cost electric vehicle (2015). https://doi.org/10.5013/IJSSST.a.16.04.03
4. S. Arof, A.K.M. Khairulzaman, J.A. Jalil, P.A. Mawby, H. Arof, Self tuning fuzzy logic controlling chopper operation of four quadrants drive DC chopper for low cost electric vehicle (2015), pp. 40–45
5. S. Arof, N.H. Diyanah, P. Mawby, H. Arof, E. Noorsal, DC drive electric car state of charge (SOC) and remaining distance traverse (RDT) optimized by fuzzy logic (2020)
6. S. Arof, N.H.N. Diyanah, N.M. Noor, M.R. Ahmad, P.A. Mawby, H. Arof, M.S. Said, A.K.M. Khairulzaman, Genetics algorithm for setting up look up table for parallel mode of new series motor four quadrants drive DC chopper (2019), pp. 155–167. https://doi.org/10.1007/978-3-030-28505-0_12
7. P. Chottiyanont, M. Konghirun, W. Lenwari, Genetic algorithm based speed controller design of IPMSM drive for EV (2014), pp. 1–5
8. S. Arof, A. Shauqee, M.R.Ahmad, N.H.N. Diyanah, P. Mawby, H. Arof, E. Noorsal, Gradient descend for setting up a look-up table of series motor four quadrants drive DC chopper in parallel mode (2021). https://doi.org/10.1007/978-3-030-67750-3_16
9. S. Arof, N.H.N. Diyanah, P.A. Mawby, H. Arof, E. Noorsal, Series motor four quadrants drive DC chopper part3: field weakening mode (2021). https://doi.org/10.1088/1757-899X/1127/1/012028
10. S. Ahmadi, S.M.T. Bathaee, A.H. Hosseinpour, Improving fuel economy and performance of a fuel-cell hybrid electric vehicle (fuel-cell, battery, and ultra-capacitor) using optimized energy management strategy. Energy Convers. Manage. **160**, 74–84 (2018)
11. B. Dianati, S. Kahourzade, A. Mahmoudi, Optimization of axial-flux induction motors for the application of electric vehicles considering driving cycles. IEEE Trans. Energy Convers. **35**, 1522–1533 (2020). https://doi.org/10.1109/TEC.2020.2976625
12. M.M. Nishat, F. Faisal, A.J. Evan, M.M. Rahaman, M.H. Sifat, Rabbi, H. M. F.: Development of genetic algorithm (GA) based optimized PID controller for stability analysis of DC-DC buck converter. J. Power Energy Eng. **8**(9), 8–19 (2020). https://doi.org/10.4236/jpee.2020.89002
13. P. Singh, A. Mahesh, GA-lookup table based firing angle optimization method for a PV connected CHB-MLI (2016), pp. 1–6. https://doi.org/10.1109/ICPEICES.2016.7853225

Chapter 14
Vehicle-To-Vehicle Data Transfer Using the Li-Fi Technology

Mohamad Rosyidi Ahmad, Nur Muhammad Tajuddin bin Abdul Manaff, Mohd Fauzi Alias, Mohamed Yusof Radzak, Saharul Arof, and Norzalina Othman

Abstract Wireless communication has become such a basic utility in our daily lives that it has become a fundamental part of our existence, and it employs the radio spectrum to convey data. Capacity, efficiency, availability, and security are among concerns that arise while using the radio spectrum. Wi-Fi use has also been shown to harm the ecology, including plants and animals. The notion of the light fidelity (Li-Fi) technology was born out of the flaws in Wi-Fi technology. Li-Fi is a sophisticated technology. This initiative is focused on vehicle-to-vehicle communication to prevent accidents on the road. Ultrasonic sensor, LDR sensor, solar panel sensor, OLED display, and Li-Fi transmitter and receiver are among the devices we utilize. If the front car experiences an abnormal situation, the vehicle in the rear will be notified and will come to a complete stop. Due to its low noise ratio, it is one of the most well-suited communication media. We use the Arduino UNO to program a laser to send binary bits with its specific ASCII code to the solar panel which then gives a character based on what binary numbers has been sent.

M. R. Ahmad (✉) · N. M. T. A. Manaff · M. F. Alias · M. Y. Radzak · S. Arof · N. Othman
Universiti Kuala Lumpur, Malaysian Spanish Institute Kulim Hi-Tech Park, Kulim, Kedah, Malaysia
e-mail: mrosyidi@unikl.edu.my

N. M. T. A. Manaff
e-mail: ntajuddin.manaff@s.unikl.edu.my

M. F. Alias
e-mail: fauzialias@unikl.edu.my

M. Y. Radzak
e-mail: myusofr@unikl.edu.my

S. Arof
e-mail: saharul@unikl.edu.my

N. Othman
e-mail: norzalina@unikl.edu.my

T. A. A. Razak et al. (eds.), *IT Application for Sustainable Living 2*,
SpringerBriefs in Applied Sciences and Technology,
https://doi.org/10.1007/978-3-031-93198-7_14

Keywords Light-dependent resistor (LDR) · Light fidelity (Li-Fi) · Universal asynchronous receiver-transmitter (UART) · American standard code for information interchange (ASCII)

14.1 Introduction

Laser communications systems are one of the methods of transferring data through the atmosphere. Laser communication operates similarly to fiber optic networks, however the beam is delivered across void space. The benefit of laser communications is that it eliminates the need for broadcast rights and cables. For laser communication to work, the transmitter and receiver must have line-of-sight conditions. Laser communications systems can be easily deployed since they are low power consumption, inexpensive, and small sized. In most cases, a laser diode is employed to produce the carrier for the transmission signal in laser communications. The basic functioning idea of Li-Fi technology is based on two concepts: transceiver and light as a medium of transmission. Li-Fi speeds are 14 Gbps when three off-the-shelf laser diodes (red, green, and blue) are utilized, and the rate is predicted to reach 100 Gbps when the entire visible spectrum is employed. According to [1], Li-Fi and VLC both employed light as a channel for data transfer. The significant difference between Li-Fi and VLC is that VLC has a unidirectional, point-to-point light communication at low data rates. Others said Li-Fi is the incorporation of Wi-Fi and VLC. [2]

Authors of [3] propose to use the Wi-Fi technology, rather than a pre-determined or manual order, to enable vehicle communication with the common traffic light system, in order to concentrate on cars and change signals accordingly. This suggested solution may appear simple to implement because traffic lights currently use LED [2] illumination. A traffic light control system receives data from an ambulance's siren lights and fire extinguishers, which switches the signal to allow for faster and uninterrupted movement. A transmitter is an electrical device that uses an antenna to emit the signal radio waves. A radio frequency alternating current is produced by the transmitter and utilized to the antenna. A photodiode module is utilized in the reception part to detect the light signal created by the Li-Fi transmitter as quickly as feasible [3].

Authors of [4] demonstrate how to use Li-Fi to transport data from one unit to a second unit. To begin, switches were used to input ASCII data into the unit, which were then saved. At the same time, this information is shown on the LCD screen. Data is encrypted and only the password is used to decode it. The data is transferred, and if the data is not received, a no response message appears on the screen.

Meanwhile, [5] have developed an external portable data storage hard-drive as an interfacing device with other secondary storage in laptop and desktop.

Fig. 14.1 Transmitter block diagram

14.2 Methodology

The system that is studied in this project is a smart system that can transmit data through Li- Fi technology of the user's front or rear vehicle. For data transfer, Li-Fi uses a laser module as a communication medium. The laser module can be used as a light source, and the 5 V solar panel will be used as a transceiver to receive and broadcast light signals. Li-Fi connects the laser module by transferring data over the light as a wireless signal receiver.

As shown in Fig. 14.1, the transmitter consists of a PC with a keyboard and an Arduino UNO. The input is given in Arduino IDE code through the keyboard (example: AbCdE). The Arduino UNO converts the text into 8 bits binary. On the transmit side, UART will generate the data packet which includes the parity bits and stop bits then transmits that packet through the transmission line with precise timing via a laser beam from the laser module.

The precaution system was added to ensure that the main system is functioning well. The threshold or analog value on the solar panel was so sensitive to the light. The analog value was set for 310, so any other value that is higher than 310 will trigger the clock signal that have been transmitted. For distance, the laser module or transmitter side need an exact distance for transmitting otherwise the solar panel would not read the clock signal. The sensor that were used are the LDR module and the ultrasonic sensor. For the LDR module, when the analog value was less than its limit, a LED will blink to alert the user about the light noise. The user will know that the system is not working now because of light interruption. The ultrasonic sensor will display to the user the distance between both transmitter and receiver before transmitting the signal. A small 5 V solar panel, an Arduino UNO, and a PC with a monitor are all found in the receiver portion. When a laser beam strikes a solar panel, it transforms the light signal carrying data into a current, and this current flows through an Arduino NANO.

14.3 Results and Discussion

Figure 14.2 shows the experimental setup. When pressing the pushbutton, the laser module beam will turn ON according to its data packet for different type of alphanumeric characters. In this project, 15 characters will be transmitted, and 15 data packets are also being created. The experiment has been conducted in different conditions. The delay time determines how fast the data will be sent. If it is low, the data can go quickly. The system was tested at distances of 1, 2, 3, and 4 m, and the output consistently displayed the right sentences and words that had been sent through the laser.

Figures 14.3, 14.4, 14.5, 14.6 and 14.7 show some examples of the data packets that have been transferred in square wave signal for each desired alphanumeric characters turning into a sentence "POTHOLES ALERT!!".

Fig. 14.2 Full hardware setup

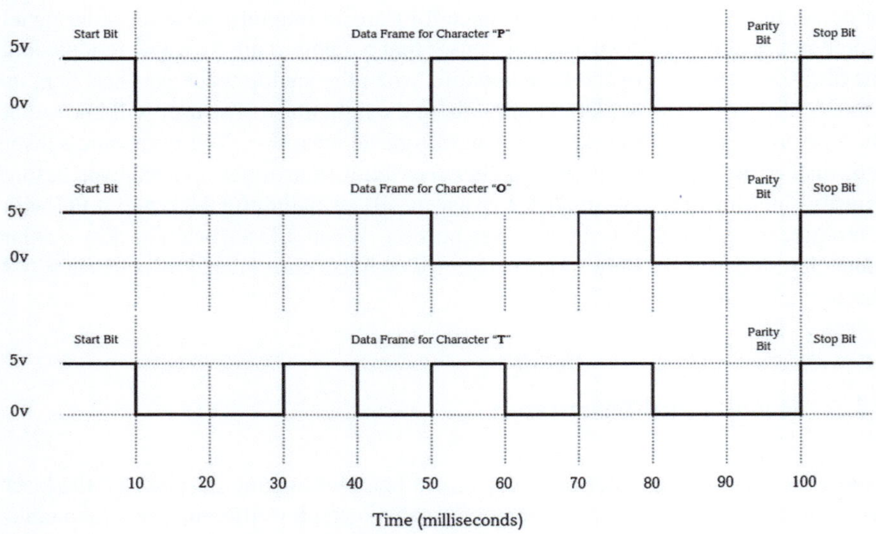

Fig. 14.3 Signal wave data packet for character "P,O,T"

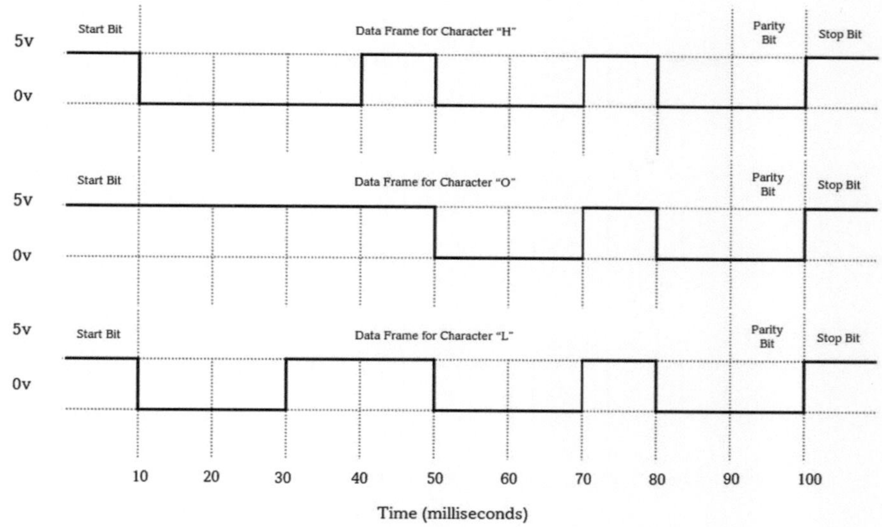

Fig. 14.4 Signal wave data packet for character "H,O,L"

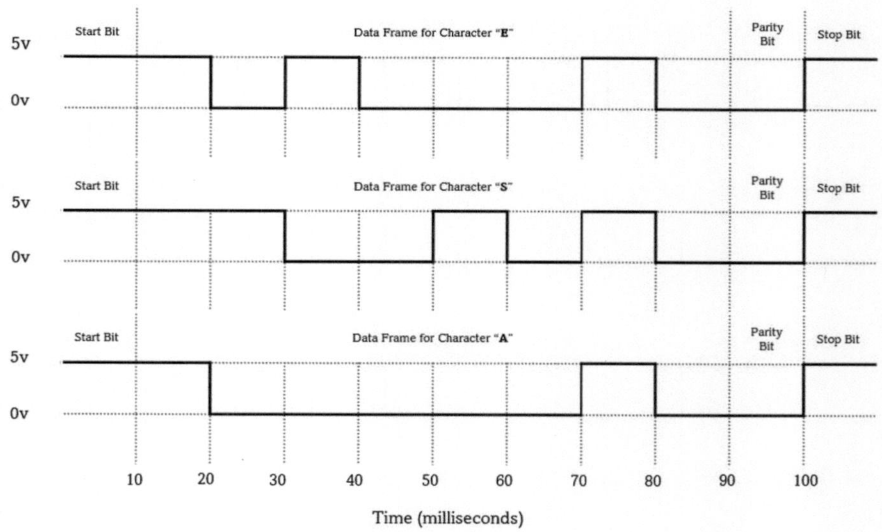

Fig. 14.5 Signal wave data packet for character "E,S,A"

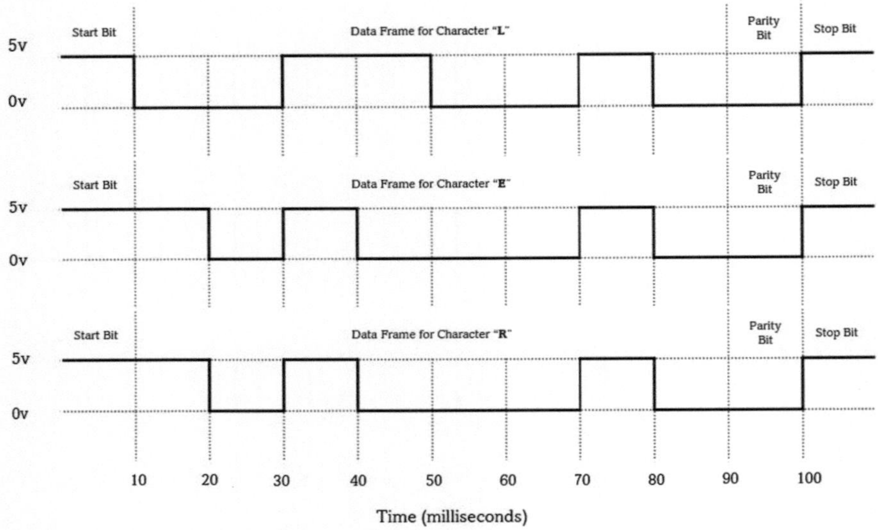

Fig. 14.6 Signal wave data packet for character "L,E,R"

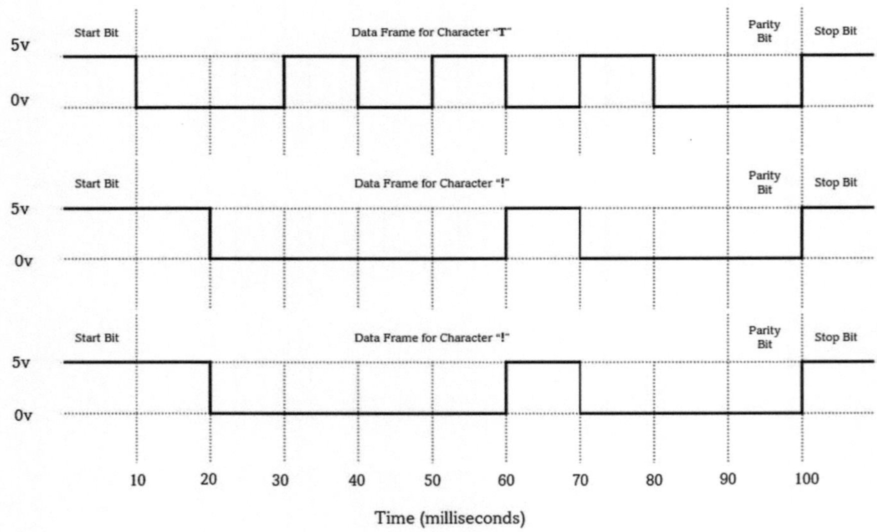

Fig. 14.7 Signal wave data packet for character "T,!,!"

(a) (b)

Fig. 14.8 The data received displayed on the OLED screen

The data sent from the transmitter will be received by the receiver. For example, the data packet sent was "POTHOLES ALERT!!" and the data was displayed on the OLED screen display shown in Fig. 14.8a. The distance of transmission and the intensity of light was also measured and is displayed on the OLED as shown in Fig. 14.8b.

14.4 Conclusion

The proposed designs in this article are not the final designs that have been implemented in the system. Before putting these plans to the test, they were constructed and assembled. There have been several changes and revisions to the basic design in the process of achieving the vehicle-to-vehicle system integration. Li-Fi is defined as a communication system, complete with asynchronous methods. The obstacles and benefits of Li-Fi are considered, with one of the main benefits being the ability to transmit data at instant speeds. Li-Fi is implemented as a V2V communication application, with data being passed from one control unit to another. The required analysis of the system with the block diagrams is also shown. It is hence determined that Li-Fi is used as a communication tool and performs well.

References

1. H. Haas, L. Yin, Y. Wang, C. Chen, What is LiFi? J. Light. Technol. **34**(6), 1533–1544 (2016). https://doi.org/10.1109/JLT.2015.2510021
2. M. Ayyash et al., Coexistence of WiFi and LiFi toward 5G: concepts, opportunities, and challenges. IEEE Commun. Mag. **54**(2), 64–71 (2016). https://doi.org/10.1109/MCOM.2016.7402263

3. S. Vijayasarathi et al., Smart traffic system using Li-Fi technology for automobiles. Int. Res. J. Eng. Tech. (IRJET) **8**(4), 534–538 (2021)
4. V. Vindal, Laser Li-Fi based SMS communication. Int. J. Emerg. Techn Eng. Res. (IJETER) **4**(4), 36–39 (2016)
5. M.Z. Faraz, A. Ambasth, Li-Fi Based Data Storage Device. Int. J. Comput. Sci. Trends Technol. (IJCST) **4**(4), 261–265 (2016)

Chapter 15
Deployment of a Canny Edge Detection Algorithm on an Embedded Device for Auto Sorting Application

Mohamed Yusof Radzak, Nurul Farhana Mohd Fadzli, Mohd Fauzi Alias, and Mohamad Rosyidi Ahmad

Abstract The automatic sorting system has been reported to be complex and a global problem. This is due to the sorting machines' failure to include flexibility in their design concept. Furthermore, products or objects may have defects and be in poor shape when they reach the hands of customers. For instance, if products on a conveyor belt are not in perfect posture or position, the robot's arm might be unable to grasp them effectively, resulting in unanticipated incidents such as objects falling or deformities due to a wrong grabbing technique. The advancement of computer vision technology is making it easier to solve this challenge with accurate results. A machine vision solution using the conventional Canny edge detector technique can be a good way to detect object deviation based on the object's edges. The algorithm is improved by incorporating the deviation angle detection technique to determine whether the objects are in the proper position for the robotic arms to grab. This research commits to developing an object deviation detection algorithm that can inspect the posture of things on the conveyor and provide information about the deviation angle. This can ensure the system's smoothness and efficacy when employed as a sorting system in industry. To evaluate this algorithm's effectiveness when applied in real time, it will be deployed to an embedded device. The NVIDIA Jetson Nano, a GPU-based device, has been selected as the target hardware. The edges and orientation angles of objects have been precisely detected by the suggested system.

Keywords Machine vision · Object deviation · Conveyor belt · Canny edge detection · Object detection

---

M. Y. Radzak (✉) · N. F. M. Fadzli · M. F. Alias · M. R. Ahmad
Electrical, Electronics and Automation Section, Universiti Kuala Lumpur, Kampus Cawangan Malaysian Spanish Institute, Kulim, Kedah, Malaysia
e-mail: myusofr@unikl.edu.my

M. F. Alias
e-mail: fauzialias@unikl.edu.my

M. R. Ahmad
e-mail: mrosyidi@unikl.edu.my

T. A. Razak et al. (eds.), *IT Application for Sustainable Living 2*,
SpringerBriefs in Applied Sciences and Technology,
https://doi.org/10.1007/978-3-031-93198-7_15

15.1 Introduction

In manufacturing, warehouses, and other locations, industrial automatics sorting, and robot manipulators are commonly used for autonomous object manipulation activities including repeated pick-and-place processes such as part picking-placing, product packing, bin picking, and kitting [1]. Typically, these stations consist of conveyor systems outfitted with vision-based monitoring systems for performing pick-and-place tasks in unstructured or dynamic environments where different objects can be mixed in arbitrary locations/positions, such as picking objects from warehouse shelves or moving conveyor belts [2]. A requirement for an algorithm that can track and calculate object poses specifically the angle of orientation of the object on the conveyor belt can ensure that the system runs properly. This paper will begin addressing the issue of object deviation on a conveyor by developing an algorithm to inspect the deviation angle. Before that, a sample of datasets was created to train the proposed algorithm and the proposed algorithm is deployed to GPU based hardware, the NVIDIA Jetson Nano.

15.2 Literature Review

Object deviation is one of the most prevalent problems in automatic sorting systems. Most of the common accidents that happen during sorting are related to the object deviation on the conveyor [2]. Due to the uneven and incorrect position of an object on the conveyor belt, accidents such as object collision have a high possibility to occur. The object can collide with each other and even worse it can cause a variety of unknown object sorting scenarios with stacking. Moreover, object deviation can lead to a wrong grasping accuracy by the robotics arm. Because of the lack of understanding of the spatial relationship of objects in the scene, the target object is damaged in the process of grasping as the objects can be falling [3].

Many existing methods generally identify the deviation of the conveyor belt, and some were doing object inspection based on features such as color, barcode, etc. However, there is no obvious way to determine the position or orientation of an object for the pick and place task [4]. To deal with this issue, this project addresses one way to identify the angle of deviation of the object on the conveyor by using Canny edge detection. The Canny edge detector is chosen as the edge detector. The angle of deviation is then calculated, and the information is used by the robotics arm making the entire sorting system more flexible. Then NVIDIA Jetson Nano will be used as a platform to deploy the algorithm and check the efficiency to be used in the automatic sorting system.

15.2.1 Edge Detection

In the research by [5], they designed the monitoring of deviation of the conveyor belt which is based on computer vision. This study develops, applies, and compares two types of conveyor condition monitoring algorithms. This research was carried out because of a highly reported deviation accident during the operation of the conveyor belt and that has become one of the most common failures. There are two approaches considered for this research.

Approach 1 is to extract the conveyor belts edges due to the Canny edge detection algorithm and the Hough line algorithm. This technology streamlines the information processing procedure after edge line extraction and achieves rapid and accurate conveyor belt status identification. In approach 2, a conveyor belt deviation detection system was created using laser technology and the LabVIEW platform. With the inclusion of the laser line, deviation detection becomes more accurate and dependable. It was shown that both approaches may perform well for belt deviation detection, particularly approach 2, which resulted in better reliability and accuracy while processing at a quicker rate.

In a previous study, the adaptive Canny edge detection algorithm approach is used to detect objects as they move along a conveyor belt in a manufacturing environment [6]. The system includes a camera to observe dark objects on a light conveyor belt. In this research, the adaptive Canny edge detection algorithm is implemented in hardware as a part of line scan edge detector. This research proved that the adaptive Canny edge detection algorithm is an effective method for edge detection via dynamic thresholds.

This proposed method is primarily determined by the changing parameters, which are the standard deviation for the Gaussian filter and its threshold values. The standard deviation determines the size of the Gaussian filter. The larger size produces a higher degree of smoothness, which is required for noisy images as well as detecting larger edges. The adaptive Canny operator detects edges with more continuity and a higher signal-to-noise ratio than other cutting-edge techniques.

15.3 Methodology /Experimental Set-Up

This research involves preparing the dataset of object deviation and testing it in the proposed algorithm to obtain the project's aim which is the object edges and the angle of deviation. The conceptual design of the algorithm focuses on detecting the object deviation on the conveyor belts in which the Canny edge detector has been chosen as the main algorithm. Extracted edges of a box will undergo deviation detection by calculating its angle of orientation. A final algorithm has been tested with the dataset and proceeded with the code generation. Deployment has been taken place to the target hardware which is the NVIDIA Jetson Nano platform. A sort of analysis was done to analyze the whole system including the accuracy of angle provided and the efficiency testing for a real time simulation.

15.3.1 Dataset

A special dataset of object deviation is created for this project to target a specific object on the conveyor belt. The scope of the objects has been reduced, and the dataset will now consist of boxes with varying positions. This study's dataset was created using actual images. A series of 30 photos captured with a mobile camera were used to train the algorithm. Each image includes objects in a range of locations and orientations with a black background. The angle extends from 0 to 360°, demonstrating the direction of the objects. The image is 960 × 1280 in size, and the distance between the objects and the camera is 40 cm. All the sample images are tested using the edge detection algorithm to detect a bad image, extreme lighting or vice versa, uneven lighting, a shadowed image, etc. Data cleaning is done to remove all the bad images (Fig. 15.1).

Fig. 15.1 Dataset of boxes with different orientation

15.3.2 Canny Edge Detector

Canny edge detection is a technique for collecting relevant structural information from a variety of visual objects while significantly lowering the amount of data to be analyzed. It is commonly employed in a wide range of computer vision systems [7]. The performance of its edge detection and positioning is good, which gives a clear response, good localization and produces only single edge response. In practical applications, the results are sometimes not satisfactory, so it is really relied on a good input image.

In this paper, the selected steps of the Canny edge detector are described as below.

Step 1: Convert original/input image to grayscale.

Step 2: Apply Gaussian filter to smooth the image to remove the noise. Figure 15.2 shows a 5 × 5 filter with variable sigma is specially used for more accuracy.

Step 3: Find the intensity gradient of the image using a Sobel filter. In this step, the vertical and horizontal gradients are calculated using convolution mask. This is fed into the magnitude computation unit, which calculates the gradient magnitude from each pixel's horizontal and vertical gradients.

Step 4: Find the magnitude, G_{mag} and angle of the gradient, θ by using Eqs. (15.1) and (15.2) respectively. Figure 15.3 shows the horizontal gradient, $G_x(x, y)$ and vertical gradient $G_y(x, y)$. These values are calculated in the x and y directions.

$$G_{\text{mag}} = \sqrt{G_x^2 + G_y^2} \tag{15.1}$$

$$\theta = \tan^{-1} \frac{G_x}{G_y} \frac{180}{\pi} \tag{15.2}$$

$$\frac{1}{159} * \begin{bmatrix} 2 & 4 & 5 & 4 & 2 \\ 4 & 9 & 12 & 9 & 4 \\ 5 & 12 & 15 & 12 & 5 \\ 4 & 9 & 12 & 9 & 4 \\ 2 & 4 & 5 & 4 & 2 \end{bmatrix}$$

Fig. 15.2 The 5 × 5 Gaussian filter

-1	0	+1
-2	0	+2
-1	0	+1

Gx

+1	+2	+1
0	0	0
-1	-2	-2

Gy

Fig. 15.3 Horizontal and vertical gradient

Step 5: Apply non-maximum suppression using interpolation to eliminate false response to edge detection. Pixels that are not part of an edge are removed during this non-maximum suppression step.

Step 6: Perform double threshold to identify probable edges and connect them.

Step 7: Perform edge tracking by hysteresis thresholding.

15.3.3 Deviation Detection

The deviation of the objects can be said to be the deflection of objects on the conveyor belt. The angle of orientation of every object is determined by comparing the top/bottom side of the objects edge with the top/bottom side of the dash box then, find the intersection, an angle was marked. In MATLAB, a function 'regionprops' with an 'orientation' property is used to perform this angle calculation. The value is expressed in degrees and ranges from $-90°$ to 90°. The angle formed by the horizontal dotted line and the major axis is called the orientation. Figure 15.4 shows the definition of correct position and an object with slight deviation.

15.4 Results and Discussion

The main objective of this research is to deploy the proposed algorithm to the targeted hardware. This was considered as an efficiency testing to test the algorithm in the real time environment and to prove that it can be implemented in a sorting system. The CSI camera attached to the Jetson Nano is used to detect objects with a resolution of 640×480. The result of the testing is shown in Table 15.1.

From the result in Table 15.1, the object is not filled with white color compared to the result during simulation in MATLAB. This is due to the resolution of the Jetson's camera and the lighting during the experiment. However, the GPU still shows the result of the angle.

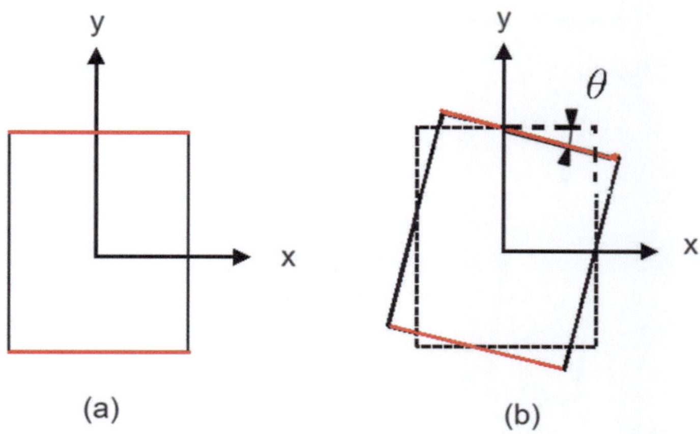

Fig. 15.4 **a** A correct position of an object on conveyor, **b** object with a slight deviation

Table 15.1 Output vision from Jetson Nano GPU

Image	Expected and actual angle (°)	Percentage error (%)
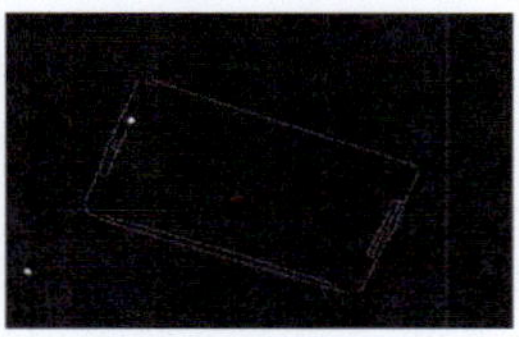	Expected angle = 72 Actual angle = 75	4
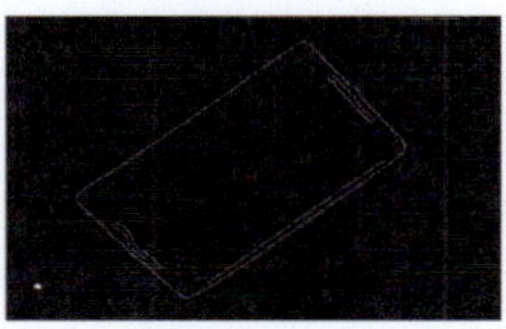	Expected angle = 49 Actual angle = 51	3.8
	Expected angle = 87 Actual angle = 88	1.1

Figure 15.5 shows the graph of the percentage error of a sample of 30 detected boxes collected during the real-time testing on NVIDIA Jetson Nano. The highest percentage error is 42% where the lowest is 0%. Most of the sample boxes produced a good angle value with a small difference between the expected angles. This real-time testing produced an acceptable result where the average percentage error is less than 10%.

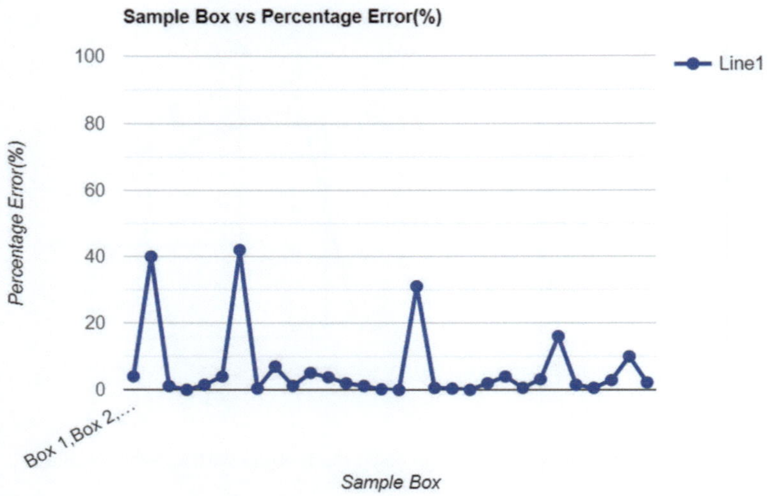

Fig. 15.5 Percentage error of 30 sample boxes

15.5 Conclusion

As a conclusion, considering the problem of deviation of objects on the conveyor belts, an object deviation detection method was proposed by taking the unique advantages of computer vision in non-contact measurement. This work evaluated the proposed algorithm for detecting the objects' deviation in an unknown scenario on the conveyor. The images of the objects in a random orientation were processed using the Canny algorithm which is deployed on the Jetson Nano GPU to extract the outer edges and an algorithm was used to determine the value of the deviation angles. The value then can be used and implemented in the robotics system in the conveyor section for correcting procedure.

References

1. S. Soltan, A. Oleinikov, M.F. Demirci, A. Shintemirov, Deep learning-based object classification and position estimation pipeline for potential use in robotized pick-and- place operations. J. Robot. **9**(3), 63 (2020). https://doi.org/10.3390/robotics9030063
2. Y. Liu, C. Miao, X. Li, G. Xu, Research on deviation detection of belt conveyor based on inspection robot and deep learning. Complexity **2021**, 1–15 (2021)
3. T. Wang, Y. Yao, Y. Chen, M. Zhang, F. Tao, H. Snoussi, Auto-sorting system toward smart factory based on deep learning for image segmentation. J. Sensors **18**(20), 8493–8501 (2018). https://doi.org/10.1109/JSEN.2018.2866943
4. K. Harada, T. Tsuji, K. Nagata, N. Yamanobe, H. Onda, Validating an object placement planner for robotic pick-and-place tasks. J. Robot. Auton. **62**(10), 1463–1477 (2014). https://doi.org/10.1016/j.robot.2014.05.014

5. M. Zhang, H. Shi, Y. Yu, M. Zhou, A computer vision-based conveyor deviation detection system. J. Appl. Sci. **10**(7), 2402 (2020). https://doi.org/10.3390/app10072402
6. F. Hossain, M. Kumar, M. Abu, Hardware design and implementation of adaptive canny edge detection algorithm. J. Comput. Appl. **124**(9), 31–38 (2018). https://doi.org/10.5120/ijca2015905446
7. A.S. Pooja, P. Smitha, Edge detection using deep learning. Int. Res. J. Eng. Technol. **5**(7), 1247–1252 (2018)

MIX
Papier aus verantwortungsvollen Quellen
Paper from responsible sources
FSC® C105338

FSC
www.fsc.org

If you have any concerns about our products,
you can contact us on
ProductSafety@springernature.com

In case Publisher is established outside the EU,
the EU authorized representative is:
Springer Nature Customer Service Center GmbH
Europaplatz 3, 69115 Heidelberg, Germany

Printed by Libri Plureos GmbH
in Hamburg, Germany